Gods of Gears

Gods of Gears

The Real Machines That Brought Ancient Myths to Life

Charles Kensington

Hidden Histories
Media

Staunton, VA 24401

Printed in the United States of America

First Printing, 2025

Published by Hidden Histories Media

Paperback ISBN: 978-1-971137-02-5
Hardback ISBN: 978-1-971137-03-2

Contents

Introduction

In a dimly lit gallery of the Museum of Ancient Technology in Athens, the air hummed with quiet reverence. A glass case glowed at the center of the room, its contents small enough to fit in the palm of a hand, a corroded gear the color of dark honey mounted on a fragment of bronze. At first glance, it looked like nothing, a relic too worn to matter. But when the curator turned a hidden crank, the ancient mechanism came alive. A series of interlocking teeth glided against one another with improbable precision, as though time itself had remembered how to move. A digital display above it lit up, showing what this device, found in the wreckage of a Greek ship more than two millennia old, once did. The gears rotated to predict lunar eclipses, planetary alignments, and the march of the zodiac. It was called the Antikythera Mechanism, and it should not exist.

Outside the museum walls, the twenty-first century whirred and blinked, filled with satellites, algorithms, and machines that think. Inside, this ancient instrument spun in perfect silence, whispering a truth both unsettling and exhilarating.

Our ancestors were far more advanced, far more inventive, than we ever dared to believe. The past, it turned out, was once the future.

The Question Beneath the Dust

The Antikythera discovery was not the first time the world realized it had forgotten its own genius. Centuries earlier, explorers in Baghdad, Istanbul, and China unearthed manuscripts describing mechanical marvels, automatic servants, musical fountains, self-moving clocks, and even steam-powered engines. Each discovery raised the same question. How could so much brilliance vanish?

This is the mystery that drives *Gods of Gears*. It is not a book about myth, but about the people who inspired myth, the engineers, philosophers, and dreamers who made the impossible move. In every culture, from Greece to Egypt, from China to the Islamic Golden Age, inventors built machines that imitated life. Their creations opened temple doors, poured wine for kings, sang to the heavens, and told time with perfect rhythm. Yet nearly all of them disappeared, their machines dismantled or forgotten, their knowledge swallowed by war and neglect.

What survived was wonder, and the quiet sense that somewhere in our history, we lost the thread that connected imagination to mechanism, faith to function, and art to science.

The Pulse of Awe

Imagine standing in a temple in Alexandria two thousand years ago. Smoke from oil lamps swirled around the heads of worshippers. A priest lit a brazier, and a deep hiss filled the air. Without a human hand touching them, the great bronze doors began to open. Fire, air, and water conspired to animate the sacred. The crowd gasped, believing they had witnessed divine power. In truth, they had witnessed something else entirely, the birth of engineering as enchantment.

That moment, when invention became indistinguishable from miracle, marked the beginning of humanity's oldest obsession, the urge to make matter move as if alive. Every civilization expressed it differently. In Greece, Hero of Alexandria built mechanical theaters where gods descended from the sky on hidden pulleys. In China, Su Song raised a tower that mirrored the heavens in perfect motion. In the Islamic world, Al-Jazari crafted automata that served wine and played music, blending geometry with artistry. Each of them sought to capture the rhythm of the universe in metal, wood, and water.

Behind their triumphs lay something more profound, a reflection of ourselves. The machines they built were not merely tools. They were mirrors, showing how deeply human the act of creation truly is.

The Lost Blueprint of the World

For centuries, the knowledge of these devices lay dormant. The blueprints that once crossed empires vanished into fire and dust. Libraries burned. Empires fell. The gears stopped turning. When later generations looked back, they saw only myths and ruins, gods that breathed, statues that spoke, and fountains that sang. The machines themselves were gone, and their magic transformed into legend.

But history has a way of resurrecting its own memory. A lost manuscript in Arabic, a corroded fragment recovered from the sea, a passage translated from Greek to Latin to English, each rediscovery added a gear to the long-forgotten mechanism of knowledge. Piece by piece, scholars began to reconstruct not just the machines but the minds that made them. What emerged was a revelation. The ancient world was not merely spiritual or philosophical. It was profoundly technological.

In temples, theaters, and royal courts, machinery expressed theology and power. Automata turned faith into spectacle and science into art. To study them is to uncover the missing chapters of human invention, the story of how the world once moved and how that motion stopped.

The Human Desire to Create Life

At its core, this is not only a story of engineering. It is a story of longing. The same impulse that drove Hero to open temple doors with fire drives modern scientists to program robots that mimic thought. The same imagination that led Al-Jazari to pour water in measured rhythm drives today's engineers to design artificial intelligence that learns. The question has never changed. Can we bring life to the lifeless?

For the ancients, this question was bound to divinity. To make a statue move was to touch the power of the gods. For modern inventors, it has become a question of intellect and ethics. Can a machine think? Can it feel? What responsibilities come with creating something that reflects us? Whether framed in myth or code, it is the same wonder, the same unease, that has followed humanity since the first gear turned.

A Journey Through Time and Imagination

Gods of Gears travels through temples and workshops, across continents and centuries, to rediscover the forgotten lineage of mechanical creation. From the hydraulic marvels of Egypt to the clockwork heavens of China, from the mechanical birds of Baghdad to the stage machines of Rome, we will trace how motion became meaning. Along the way, we will meet the minds who bridged science and soul, engineers who saw no boundary between invention and art, between physics and faith.

This journey is both archaeological and emotional. It digs into ruins and manuscripts, but it also explores the inner machinery of curiosity, the quiet force that compels us to understand and to create. Every turning wheel and dripping clepsydra in these pages is a symbol of that force. Together, they form the pulse of human history, the rhythm that still drives us to move, to build, to dream.

The Echo of Motion

The museum lights dimmed. The curator closed the case around the Antikythera Mechanism. Its gears came to rest, but the echo of their motion lingered. Visitors walked away, unaware that they had just witnessed something miraculous, the past remembering itself.

Outside, the modern world hummed with its own miracles, satellites orbiting Earth with clockwork precision, prosthetic limbs that moved at a thought, artificial voices that spoke from invisible circuits. Yet the essence was unchanged. We were still the same species that once stood in a smoky temple and gasped when a statue seemed to breathe.

The machines have returned, refined and multiplied, but their origin lies deep in the ancient imagination, a place where invention and divinity met and where motion first became a language of wonder.

As you turn the page into Chapter One, step into that world again. Stand in the temple's flickering light. Hear the gears begin to turn. The gods are moving, and the story of how they came to life is about to begin.

Chapter 1
The Gods That Moved — Temples, Illusion, and the Birth of Ancient Automation

Smoke curled upward from the altar, thick with the scent of myrrh and oil. The dim temple glowed in the wavering light of torches as the faithful knelt, their faces half lit and reverent. The priest raised his arms toward the great bronze doors that sealed the inner sanctum. For a heartbeat, nothing stirred. Then a low rumble echoed through the chamber. The torches flared brighter. The air grew warmer. Slowly, the massive doors groaned open by themselves. The congregation gasped, overcome with awe. To them, it was proof that the god had heard their prayers. To the hidden engineer who designed the mechanism, it was the elegant obedience of physics to human will.

This was not fantasy. It was the earliest expression of automation, an art that blended science, faith, and performance to create the illusion of the divine made visible. Centuries before machines served industry, they served wonder. In the temples of Greece and Egypt, mechanisms of air, fire, and water moved statues, opened doors, and poured libations as if animated by unseen gods. The true miracle was not magic, but mastery.

The Power of Sacred Motion

In the ancient world, motion was a language of the divine. The Greeks and Egyptians believed that life was breath, that the act of moving or speaking set the gods apart from inert matter. To make stone move or metal act with purpose was to imitate creation itself. Engineers, known variously as mechanicians, inventors, or philosophers, became the unseen interpreters of divine will. Their craft made worship a spectacle.

Temple automation arose not from idle curiosity but from devotion. Religious ritual was the center of public life, and every act of worship was carefully staged to inspire emotion. The more vivid the ritual, the deeper the faith. Priests sought ways to make gods appear responsive, to make an altar flame flicker as a prayer ended, to make a statue pour wine after an offering. The effect had to be seamless. The worshippers could never see the pipes, weights, or pulleys hidden behind the marble walls.

The idea of manipulating nature's elements for ritual was rooted in centuries of observation. Ancient engineers knew that heat caused air to expand, that air pressure could move water, and that gravity could set motion in sequence. By combining these forces, they could choreograph a mechanical drama in which the divine seemed to participate.

Egypt: The First Engineers of Faith

Long before Hero of Alexandria recorded his designs, Egypt had already become a cradle of mechanical wonder. Along the Nile, engineers had perfected the shaduf and water wheel, devices that harnessed balance and flow. These same principles found their way into temples, where sacred architecture and subtle hydraulics converged.

Inscriptions and archaeological evidence suggest that certain Egyptian sanctuaries, particularly those dedicated to Amun Ra, used hidden mechanisms to heighten ritual experience. Doorways operated by counterweights could open at key moments during a ceremony. Hidden channels beneath altars carried water that seemed to flow spontaneously, symbolizing purification.

Some accounts even describe statues that breathed or spoke. The most famous example is the Colossus of Memnon near Thebes. Each morning, as the rising sun warmed its stone, a tone resonated from deep within the monument. Ancient travelers described it as the voice of a god greeting the dawn. Modern analysis suggests the sound came from expanding air trapped in cracks within the structure, a natural effect mistaken for divinity. Yet the fascination it inspired shows how eagerly people embraced the idea of speaking monuments and living idols.

Priests guarded these secrets carefully. Their role depended on their ability to mediate between mortals and gods, and mechanical marvels strengthened that illusion. When the pharaoh entered a temple and the doors swung open without human touch, it was not technology, it was power made manifest. The temple itself became a living organism, animated by forces only the priesthood could command.

Hero of Alexandria: The Mind Behind the Miracles

By the first century CE, Egypt was under Roman rule, and the city of Alexandria stood as a crossroads of science and faith. In its libraries and workshops, Greek mathematics merged with Egyptian practicality. Here lived Hero of Alexandria, the man who transformed scattered temple tricks into a system of mechanical principles.

Hero's treatise *Pneumatica* remains one of the most extraordinary technical works of antiquity. In it, he described dozens of machines powered by air, water, and fire. His purpose was not to impress scholars but to explain how the unseen could be made to move the visible. His tone is methodical, as though describing a geometry problem, yet behind his equations lies an unmistakable sense of wonder.

The most celebrated of his inventions was his self opening temple door, a masterpiece of engineering theater. A fire was lit on the altar above a sealed bronze sphere partly filled with air. As the heat increased, the air expanded and forced water through a narrow pipe into a suspended bucket. The added weight pulled a chain wound around a drum, which in turn drew open the temple doors. When the priest extinguished the flame, the air cooled and contracted, the water flowed back into the sphere, and the doors swung shut again.

To the congregation, the gods had answered. To Hero, it was proof that nature itself could perform miracles under guidance of reason.

The Beauty of Balance

Many of Hero's other devices played upon the same principle, the invisible transformation of pressure into motion. He designed fountains that flowed when torches were lit, statues that poured libations in timed sequences, and birds that sang through hidden flutes when air passed through their cavities. His rotating theatrical automata, powered by simple weights and pulleys, enacted miniature dramas that rivaled any modern mechanical stage.

Hero's fascination with pneumatics, the breath of air, was not accidental. In Greek thought, pneuma was both a physical and spiritual concept, the breath of life. To make air act was to command the essence of animation itself. Each device, no matter how simple, carried symbolic meaning, life within the lifeless.

The Hidden Theater of Faith

The genius of temple automation lay not just in its mechanics but in its secrecy. The worshippers saw miracles; only a handful of initiates knew the truth. Temples were designed to conceal machinery within thick walls, columns, or platforms. Pipes disguised as decorative features carried air or water. Weighted pulleys hid inside altars.

Rituals were timed so that each mechanism activated at the perfect emotional moment. When the priest prayed for the god to open the temple, the fire would already be lit, and the expanding air would begin its work. By the time the prayer reached its climax, the doors would move. Faith and physics synchronized in careful harmony.

This staging was not deceit. It was devotion. The ancients believed that understanding nature's laws was itself a divine pursuit. By revealing the connection between the elements, engineers honored the gods through imitation. To create movement from stillness was an act of reverence, not rebellion.

Faith, Authority, and the Machine

These marvels also served political power. The connection between priestly authority and mechanical spectacle was profound. When a ruler or high priest commanded the gods to act, and the gods obeyed, their legitimacy was reinforced. The worshipper's awe became submission.

In Greece, as in Egypt, temples were not only places of worship. They were symbols of control. Automata transformed ritual into demonstration. Every hidden valve and lever declared that the priesthood understood the forces of the universe better than ordinary men. This marriage of knowledge and faith echoed through centuries, influencing later empires and even medieval church architecture, where organs and clockwork figures continued the tradition of sacred motion.

Evidence from the Ancient Record

While Hero's *Pneumatica* survives as the clearest window into this lost art, traces appear elsewhere. The works of Philo of Byzantium, Ctesibius, and Vitruvius describe devices that predate Hero's own. Ctesibius, often called the father of pneumatics, invented water clocks with moving figures and pressure driven pumps centuries earlier. Vitruvius, in his *De Architectura*, mentioned similar contraptions designed to impress worshippers.

Fragments of bronze tubing and ceramic piping found at temple sites support these accounts. Scholars have reconstructed several of Hero's mechanisms using materials described in his texts, confirming their practicality. When replicated, the devices perform exactly as he claimed. Watching a modern reconstruction of Hero's self opening doors is as captivating now as it must have been two thousand years ago. The doors swing open slowly, silently, as the altar fire burns above them.

The Legacy of Motion

What Hero and his predecessors created was more than mechanical curiosity. It was the foundation of a philosophy. By merging observation, mathematics, and artistry, they revealed that natural forces could be directed by intellect. Their work bridged two worlds, the divine and the rational.

The wonder of those early temple devices still resonates today. In every automatic door and every mechanical display echoes the same impulse that drove Hero and the priest engineers of Egypt, the desire to give motion to the inert, to make human creation imitate divine vitality.

When the final torch dimmed in a Greek temple and the doors shut once more, the worshippers departed believing they had seen the hand of the gods. Yet behind the curtain of incense and stone, an unseen engineer smiled in quiet triumph, knowing the miracle had not descended from heaven but risen from human ingenuity.

Chapter 2
The Engineer of Miracles – Hero of Alexandria and His Mechanical Legacy

The small bronze bird sang its metallic song, a sound so pure it seemed alive. The apprentice jumped, half laughing and half afraid, while Hero of Alexandria adjusted the copper tubing with practiced precision. Steam curled faintly from the vessel beside him, carrying the sharp scent of heated oil. Around him, his workshop was alive with motion, spinning pulleys, dripping valves, weights descending in measured rhythm. It was a room filled not with mystery, but with mastery. Here, in this dim workspace overlooking the great port of Alexandria, Hero transformed the laws of nature into living art.

The Heart of a Hellenistic Genius

Hero lived at a crossroads of worlds. By the time of his birth in the first century CE, Alexandria had become the intellectual capital of the Mediterranean, a place where Egyptian engineering, Greek mathematics, and Roman ambition converged. Founded by Alexander the Great and nurtured by the Ptolemies, the city pulsed with invention. Its towering lighthouse, the Pharos, guided ships from miles away; its vast library preserved the thoughts of the world's greatest minds. Hero inherited that legacy.

A mathematician, physicist, and inventor, Hero represented the last great spark of Hellenistic science before it dimmed into the long centuries of decline that followed Rome's expansion. Unlike philosophers who speculated on abstract principles, Hero concerned himself with the tangible. He asked how fire, air, and water could be bent to human will. His answers came in the form of devices that moved, sang, poured, and responded.

To later generations, he would seem almost mythical, a man who bridged the gap between the ancient dream of animation and the modern reality of engineering.

The Workshop of Wonders

Hero's surviving writings reveal not only a brilliant mind but an extraordinary teacher. His *Pneumatica*, *Automata*, and *Mechanics* describe dozens of machines so precisely that modern engineers have recreated them. In these pages, one can almost hear the hiss of air and the click of gears as he explains, step by step, how to bring motion from stillness.

His *Automata* describes a miniature theater, a complex stage powered by ropes and weights concealed within its base. When set in motion, it performed an entire play, complete with moving figures, shifting scenery, and sound effects, without human touch. The action unfolded through a carefully timed system of pulleys that released each motion in sequence. The curtain rose, actors advanced, a thunderclap echoed, and finally the curtain fell, all governed by mechanics.

It was entertainment, yes, but it was also philosophy in motion. To Hero, such creations demonstrated the harmony of mathematics and the natural world. He showed that precision could mimic life, that beauty could be engineered.

The Coin That Bought the Miracle

One of Hero's most remarkable designs was a coin operated holy water dispenser. The principle was simple yet astonishingly modern. When a worshipper placed a coin into the slot, it fell onto a small pan attached to a lever. The weight of the coin tilted the pan, opening a valve and allowing a measured flow of water to pour from the spout. When the coin slid off, the lever rose again and the valve closed.

The transaction was symbolic. In the temple, offerings were exchanged for blessings; Hero's machine made that exchange literal. It also introduced a principle that would reappear two thousand years later in vending machines and automated kiosks: a self regulating system that responded predictably to human input. Hero had turned faith into feedback, and ritual into mechanism.

The Fire That Moved Doors

Hero's temple doors, described in *Pneumatica*, embodied his genius for uniting science and spectacle. The mechanism, powered by fire and air pressure, gave the impression that the gods themselves answered the priest's command. Hidden beneath the altar, a sealed vessel expanded as it heated, forcing water through a pipe and into a suspended bucket. The bucket's weight pulled a chain that opened the doors. When the fire was extinguished, the process reversed, and the doors closed once more.

This design blended physics with theater. It was not enough for Hero to make a device that worked; it had to mean something. The altar's flame symbolized divine presence, so when that flame caused the temple to move, it reinforced belief. In Hero's hands, science did not replace faith. It gave it motion.

The Aeolipile: The Breath of the Gods

Perhaps the most famous of Hero's creations was the Aeolipile, often called the first steam engine. It consisted of a hollow metal sphere mounted on an axis, with two bent tubes protruding from its sides. When water in the sphere was heated, steam escaped through the tubes, causing the sphere to spin.

It was a demonstration, not a tool, an illustration of how expanding vapor could produce motion. Yet its significance cannot be overstated. The same principle would one day power locomotives, turbines, and entire civilizations. Hero's device was small, simple, and almost playful, but it marked a turning point in human understanding of energy. The ancients had harnessed the wind and the river; Hero showed that even air and heat could be tamed.

The Philosopher of Motion

Though Hero is remembered as an engineer, he was also a philosopher of practicality. He believed that knowledge should serve demonstration, that ideas gained meaning only when proven through mechanism. His devices were not curiosities but arguments, evidence that the universe was orderly, rational, and open to human mastery.

In his mathematical works, Hero devised methods still taught today. His formula for calculating the area of a triangle from its sides, known as Hero's Formula, remains a cornerstone of geometry. His writings on optics and mirrors reveal an understanding of reflection that prefigured later discoveries in physics.

Yet, despite his brilliance, Hero lived at the twilight of an age. The scientific culture that had flourished since Aristotle and Archimedes was fading. Alexandria's libraries faced decline; political turmoil weakened royal patronage. The great workshops that once built mechanical theaters and water organs fell silent. Hero's inventions would outlive him only through the survival of his manuscripts.

A Bridge Across Centuries

Hero's influence did not end with Rome. Centuries after his death, his works resurfaced in the Islamic Golden Age, preserved and translated by scholars in

Baghdad, Damascus, and Cairo. The *Pneumatica* was rendered into Arabic as *Kitab al Hiyal*, "The Book of Ingenious Devices." There it inspired new generations of inventors who saw in Hero's principles the blueprint for their own wonders.

The Banū Mūsā brothers, ninth century engineers working under the Abbasid caliphs, expanded on Hero's designs, creating musical fountains, programmable servants, and mechanical animals. Their treatise, *The Book of Ingenious Devices*, cites earlier Greek sources, including Hero. A century later, Al Jazari carried the tradition even further, building intricate clocks and automata that delighted the courts of the Islamic world.

Through these scholars, Hero's legacy traveled westward once more. Medieval Europe rediscovered his works through Arabic translations. By the Renaissance, his texts were printed and studied by scientists eager to revive the mechanical arts. Leonardo da Vinci, fascinated by automata and hydraulic motion, likely encountered references to Hero's machines. The lineage of thought, from the bronze bird in Hero's workshop to the clockwork knights of Renaissance Italy, is direct and enduring.

The Last Light of Alexandria

Hero's life unfolded as Alexandria's own glory waned. The fires that once illuminated the library dimmed under Roman control, and the city's role as the world's scientific heart gradually eroded. Yet in Hero's work, the flame endured. His machines embodied a principle that would reemerge centuries later: that knowledge could imitate nature, and imitation could reveal truth.

In the flickering light of his workshop, surrounded by tools and scrolls, Hero bridged the ancient and the modern. He proved that invention was not mere curiosity. It was communication, a dialogue between human reason and the universe itself.

When he set the copper bird to sing, he demonstrated more than sound. He showed that the world could be written in motion, that even breath and flame could become language in the hands of one who understood. Through him, the dream of animation passed from myth into mechanism, from the gods' domain into humankind's grasp.

Hero of Alexandria did not simply build machines. He built a future.

Chapter 3
The Festival of Motion — Greek Theaters and the Machinery of Storytelling

The audience held its breath as the chorus fell silent. From behind a painted backdrop, a creak echoed, then the air seemed to shift as a figure appeared above the stage. Suspended by unseen ropes, a gleaming wooden god descended from the heavens, light catching on his gilded robes. Below, the actors froze in reverence. The crowd gasped. This was the moment they had waited for, the divine arrival, the deus ex machina, literally "god from the machine."

The spectators in Athens' great theater at the foot of the Acropolis were not simply watching a play. They were witnessing the birth of mechanical storytelling. Beneath the sun, in amphitheaters carved from stone, Greek drama became a union of art, architecture, and engineering. Every pulley that turned and every platform that revolved served a single purpose, to make the invisible visible, to bring myth to life before human eyes.

The Origins of Theatrical Mechanics

The roots of Greek stage machinery stretch back to the fifth century BCE, during the Golden Age of Athens. Theaters had already evolved from simple wooden platforms into vast semicircular arenas that could hold thousands. These open air structures, known as theatron, "the seeing place," were both civic gathering spots and sacred spaces dedicated to Dionysus, the god of wine, ecstasy, and drama.

The earliest plays required little more than masks, robes, and imagination. But as tragedy matured into the state's most honored art form, the demand for spectacle grew. Audiences expected transformation, mortals becoming gods, ships appearing on the horizon, thunder rolling from the sky. The playwrights Aeschylus, Sophocles, and Euripides wrote scenes that defied the limits of human performance. To meet these challenges, the Greeks invented mechanical illusion.

At the heart of this transformation were devices that blended physics and pageantry. The stage became a laboratory of levers, cranes, and rotating scenery. What began as simple aids to storytelling evolved into a complex system of machinery that would influence theater design for two millennia.

The Crane That Lowered the Gods

Perhaps the most famous of all theatrical machines was the mekhane, the crane that literally brought gods to earth. Constructed of wooden beams and operated by ropes and pulleys, the mekhane allowed actors playing divine beings to be lifted above the stage or lowered dramatically from the heavens.

The operator, hidden behind the backdrop, used counterweights to control the descent. To the audience, the movement appeared effortless, a sudden celestial appearance accompanied by music or chanting. The mekhane transformed the stage into a bridge between worlds, embodying the Greek belief that the gods watched and intervened in mortal affairs.

This device gave rise to one of drama's most enduring concepts, the deus ex machina. When playwrights used divine intervention to resolve impossible plots, it was often literal: a god, suspended on the crane, would descend to deliver judgment or mercy. Though modern critics often use the term to describe a contrived ending, for the Greeks it expressed divine authority and cosmic balance.

The Rotating Platforms of Fate

Another crucial invention was the ekkyklema, a wheeled platform that could be rolled out from behind the skene, or stage building, to reveal scenes that took place offstage. Greek sensibilities forbade the depiction of violence in full view of the audience, so the ekkyklema served to display its aftermath. After a hero's death or a battle's end, the platform would roll forward, showing the fallen figure or the destruction wrought, a silent tableau of consequence.

The ekkyklema was more than a convenience. It was a moral instrument. By controlling what the audience could and could not see, it framed tragedy as reflection rather than spectacle. The machine turned narrative into ritual, reinforcing the Greek conviction that human suffering was both inevitable and instructive.

The Periaktoi: Turning Worlds

To change scenes, Greek engineers developed the periaktoi, tall triangular prisms that could be rotated to reveal different painted faces. Each side displayed a different backdrop, a palace, a forest, the sea. With a turn of the mechanism, the world shifted.

This simple but ingenious device introduced the idea of cinematic transition long before the invention of film. The periaktoi allowed stories to move swiftly through space and time without breaking the audience's immersion. Theatrical imagination found its mechanical ally in geometry.

The Sound and Fury of the Stage

The Greek theater appealed to all the senses. Beneath the wooden stage, hidden technicians operated devices that mimicked thunder and wind. Bronze sheets were shaken to produce rumbling. Hollow jars buried in the theater's structure amplified resonance, an early experiment in acoustic engineering.

These techniques were not merely tricks. They were integral to the religious nature of performance. Theater in ancient Greece was an act of worship. The festivals of Dionysus were civic events that unified the city-state through shared emotion, fear, pity, awe. The mechanical illusions served to heighten these feelings, making divine justice and human frailty palpable.

When the thunder rolled and Zeus appeared on the mekhane, the boundary between myth and reality dissolved. For a few moments, mortals could glimpse eternity.

The Collaboration of Art and Science

Behind the grandeur of the theater stood craftsmen whose names are mostly lost. Carpenters, architects, and mathematicians worked together to ensure that every device functioned flawlessly under the Mediterranean sun. The precision required was immense. The ropes of the mekhane had to bear the weight of a grown man while moving smoothly enough to maintain the illusion of flight. The ekkyklema needed to roll quietly and evenly across uneven stone.

These early stage engineers were the precursors to later mechanical inventors like Hero of Alexandria. Indeed, Hero's writings mention similar pulley systems and counterweights, showing how theatrical ingenuity fed into temple mechanics and beyond. The theater, the temple, and the workshop shared a common ancestry. All sought to animate the inanimate.

The close relationship between engineering and performance reflected a deeper Greek ideal, techne, the unity of art and craft. To the Greeks, a skilled artisan was not lesser than a philosopher. Both pursued truth through creation. The stage, with its balance of structure and surprise, embodied that ideal.

The Cultural Power of Spectacle

Greek audiences came to the theater expecting not only storytelling but transformation. The plays mirrored their society's struggles, war, pride, fate, but elevated them through divine perspective. The presence of mechanical gods reminded viewers that mortal life was small within the vast order of the cosmos.

This spectacle had a moral dimension. The gods who descended at the end of a play did not exist to entertain; they restored balance. The deus ex machina reaffirmed that justice, divine or otherwise, would prevail. Even the machinery that made it possible carried symbolic weight. The ropes, levers, and pulleys mirrored the invisible forces that governed both stage and life.

The festivals themselves, especially the City Dionysia in Athens, were celebrations of civic identity. Thousands gathered in the theater not as individuals but as citizens, witnessing stories that bound their community through shared catharsis. When the machinery worked perfectly, it became invisible. Only the emotion remained.

Engineering the Invisible

The Greeks approached stagecraft with the same reverence they gave to geometry and architecture. They understood proportion, balance, and rhythm not as artistic flourishes but as laws of the universe. The machines of the theater obeyed those same laws. Each lever and counterweight embodied harmony.

The perfection of these devices required rigorous experimentation. Craftsmen tested the strength of ropes, the smoothness of axles, and the reliability of pivot points. Failures could be disastrous. An ill timed release might send an actor tumbling instead of flying. Yet the persistence of these inventions across centuries shows their reliability. Roman theaters later adopted and expanded on the same designs, adding trapdoors, elevators, and elaborate scenery changes, but the Greek foundations remained.

The Legacy of Theatrical Motion

When audiences left the amphitheater, they carried with them more than memories of performance. They had witnessed a merging of art, religion, and technology that few other cultures would achieve for centuries. The mechanical gods who descended on ropes foreshadowed the automatons of Hero's temple doors and the animated statues of medieval cathedrals. The theater became the prototype for the mechanical imagination itself.

In time, the language of motion migrated from the stage to the workshop, from performance to invention. The principles of balance, timing, and hidden force inspired generations of engineers long after the last chorus faded.

The Greek theater was, in truth, the first great school of applied mechanics. Within its stone walls, human beings learned to command illusion, to choreograph wonder. Every pulley and counterweight was a lesson in the divine geometry of creation.

When the final god descended, bringing resolution to the tragedy below, the crowd rose in reverent applause. The ropes were hidden, the levers silent. The divine had spoken through mechanism, and for that fleeting moment, the line between faith and engineering vanished into pure astonishment.

Chapter 4
The Bronze Man of Crete — Talos and the Line Between Myth and Machine

Waves crashed against the rocky coast of Crete, sending spray into the air like shards of glass. On the horizon, a figure moved, massive, gleaming, and unmistakably human. As the sun struck its bronze limbs, light scattered in dazzling flashes across the sea. The sailors, weary from days at sea, froze in terror. The figure raised an arm and began to run along the shoreline, its footsteps shaking the ground. This was no man of flesh and bone. This was Talos, the bronze guardian of Crete, and he would allow no intruder to land.

So the story goes. The Greeks told of Talos as a living statue, forged by the god Hephaestus and given to King Minos to protect the island from invasion. He circled Crete three times each day, an unceasing sentinel made of metal yet filled with life. His single vein, sealed with a bronze nail, carried the divine ichor that animated him. To later generations, Talos became both marvel and mystery, a being that blurred the boundary between mythology and machinery.

The Legend of the Living Metal

Ancient poets delighted in describing Talos's impossible vitality. Hesiod, Apollonius of Rhodes, and later writers painted him as the ultimate guardian, tireless, radiant, and deadly. When foreign ships approached, Talos hurled stones from the cliffs or heated his bronze body until it glowed, then embraced intruders in a searing death. His demise came when Medea, the sorceress who traveled with Jason and the Argonauts, tricked him into removing the nail that sealed his lifeblood. As the ichor poured out, the giant collapsed, a machine undone by its own design.

To modern ears, the tale sounds like fantasy. Yet to the Greeks, Talos was not just myth. He represented something both wondrous and plausible. They lived in a world already enchanted by mechanical marvels, singing birds, self-opening doors, and statues that moved. Talos was the divine exaggeration of an idea already being explored in temples and workshops, the dream of giving motion to matter.

Forged by Gods, Inspired by Men

The god credited with Talos's creation, Hephaestus, was no ordinary deity. In myth, he was the blacksmith of Olympus, the maker of automata. Ancient texts describe him forging self-moving tripods that rolled on golden wheels to serve the gods during banquets. In Homer's Iliad, Hephaestus commands mechanical maidens crafted from gold and endowed with intelligence, speech, and strength. These stories, though wrapped in divine language, reveal the Greeks' early fascination with artificial life.

By assigning these creations to Hephaestus, the ancients acknowledged the mystery of technology. To them, such marvels belonged to the realm of gods and heroes, yet they also reflected the experiments of real artisans who worked bronze and iron in smoky forges across the Aegean. The line between divine invention and human craft was thin. When mythic storytellers spoke of metal that moved, they echoed the tangible power of metal shaped by fire.

The Reality of Bronze and the Minoan Imagination

Long before classical Greece, the island of Crete had been home to the Minoan civilization, one of the earliest advanced societies in Europe. Flourishing between 2000 and 1400 BCE, the Minoans were master metalworkers. Archaeological finds from Knossos and other palatial centers show remarkable skill in bronze casting: statues, blades, mirrors, and delicate ritual vessels created with sophistication that astonished later cultures.

Bronze was more than a material; it was a symbol of permanence. Unlike wood or clay, it endured. When heated in a forge, it seemed to transform, solid one moment and molten the next. To Minoan eyes, such metamorphosis must have seemed magical, an echo of divine fire. It is not difficult to imagine how stories of living metal took root in a culture that witnessed lifeless ore become gleaming form.

Minoan artisans may not have built machines like Hero of Alexandria's centuries later, but they understood balance, hydraulics, and motion in other ways. The famed Minoan plumbing systems at Knossos, with their pressurized clay pipes and sophisticated drainage, reveal a mastery of controlled flow, an early echo of the principles behind pneumatic engineering.

If temple engineers of the Hellenistic world gave birth to mechanical automata, it was the Minoans who first gave metal its mystique. In their art, figures of dancers, bulls, and deities appear poised in motion, capturing energy within stillness. Talos, born from this heritage, became the mythic embodiment of a culture that already believed matter could be animated.

Between Myth and Mechanism

Greek thinkers never dismissed tales like that of Talos as mere fantasy. Instead, they debated their meaning. Was the bronze man a metaphor for divine power or a symbol of human invention? Philosophers such as Plato and Aristotle grappled with questions of motion, soul, and imitation. If a craftsman could create an object that moved on its own, did it possess life, or only the illusion of life?

Aristotle's concept of entelechy, the inner drive that makes a thing fulfill its purpose, echoes faintly in the myth of Talos. The bronze guardian acted with intent, patrolling his island tirelessly and fulfilling his function without deviation. To the philosophers of the ancient world, that raised unsettling possibilities. If motion could be created without consciousness, then what separated human will from mechanical cause?

The Greeks embraced the paradox. Talos was both tool and creature, statue and soldier. His heart was a forge, his body a work of art, his mind an enigma. By embodying the tension between creation and autonomy, he became the earliest metaphor for the modern machine.

The Art of the Living Statue

Accounts of self-moving statues persisted throughout Greek history. In the second century CE, the traveler Pausanias wrote of temple images said to move or speak during rituals. Some were described as nodding their heads in response to prayers, others as lifting arms or opening eyes. Many historians interpret these as mechanical illusions powered by concealed pulleys or air pressure, devices like those Hero of Alexandria would later perfect.

These animated statues served the same purpose as Talos: to make the divine appear imminent. The act of movement transformed sculpture into presence. Worshippers no longer beheld a symbol of the god; they stood before what felt alive. In this sense, every self-moving statue was a descendant of Talos, every hiss of air through a hidden pipe an echo of his bronze breath.

From Forge to Faith

Bronze carried spiritual weight. The Greeks believed it could channel divine energy, its reflective surface a mirror to the heavens. Many votive figures and ritual vessels were cast from bronze not only for durability but for its luminous quality, the way it caught firelight and seemed to glow from within. To an ancient mind, this shimmer hinted at life.

When myth told of a bronze man animated by ichor, it extended the logic of the forge. Heat and motion already turned raw metal into creation. If fire could awaken matter once, why not again? The myth of Talos took this process to its divine extreme.

In some versions of the tale, Hephaestus created Talos not as a weapon but as a teacher, a symbol of technology's dual nature. The same fire that shaped civilization could also destroy it. Talos guarded Crete with relentless devotion, yet his strength isolated him. His collapse served as a warning: even the most perfect creation carries within it the flaw of its maker.

The Legacy of Talos

Over centuries, Talos evolved from myth into metaphor. To Hellenistic thinkers, he represented the union of art and science. To poets, he was the tragic embodiment of human ambition. The story survived well into the Roman period, where it fascinated scholars and dramatists. In later centuries, when automata reappeared in Islamic and European engineering, Talos returned as inspiration, the first imagined machine that could walk, fight, and die.

Today, historians view him as the symbolic ancestor of robotics. His legend captures the same curiosity that drives modern engineers to build machines that imitate life. Every artificial intelligence and every humanoid automaton carries a trace of Talos's bronze shadow.

The Shining Sentinel

As the sun sets over Crete's jagged coast, the waves still break against the rocks where Talos was said to fall. The sea glows copper under the fading light, as if remembering its ancient guardian. To stand there is to feel the same awe that once filled the sailors' hearts, the wonder of motion where none should exist, the uneasy admiration for a creation too perfect and too alive.

In that twilight between myth and mechanism, Talos endures, part god, part machine, and entirely human in the questions he leaves behind.

Chapter 5
The Clockwork Dragon – Su Song's Tower of Heaven

In the heart of 11th century Kaifeng, under the Song dynasty's brilliant skies, a tower rose higher than the city walls, a structure that shimmered with bronze and purpose. Inside, hidden from the bustling streets, a great mechanism breathed in rhythmic precision. Wooden gears meshed with bronze teeth. Chains clinked in deliberate intervals. Above, armillary spheres traced the movement of the stars. Every rotation echoed the heavens themselves. This was the water driven astronomical clock tower of Su Song, a triumph of ingenuity so extraordinary that the Chinese called it the Tower of Heaven.

The tower stood not only as a marvel of engineering but as a declaration of divine order. In its harmonious motion, time and the cosmos converged, expressing a truth the Song emperors cherished, that their rule was aligned with the universe itself.

The Scholar of Many Worlds

Su Song was no mere craftsman. Born in 1020 CE, he was a statesman, diplomat, astronomer, pharmacologist, and polymath whose curiosity spanned the realms of science and governance alike. Educated in the Confucian classics, he rose to serve the imperial court as Minister of Personnel, later directing the Bureau of Astronomy. His intellect was matched only by his ambition to design a machine that could embody the harmony between Heaven and Earth.

In Chinese cosmology, the emperor ruled by virtue of the Mandate of Heaven, a moral and cosmic principle that connected good governance with celestial balance. When Su Song proposed constructing a great clock tower to track the motions of the stars, planets, and time itself, he was not merely pursuing scientific discovery. He was reinforcing the divine legitimacy of the throne.

A Machine to Imitate the Cosmos

The clock tower stood nearly thirty feet high, divided into multiple levels, each representing a layer of the cosmos. At its heart was a water powered mechanism unlike anything the world had seen before. Earlier timekeeping devices in China had used dripping water to mark hours, but Su Song transformed the principle into a dynamic, self regulating system.

At the base, a massive bronze clepsydra, a water reservoir, fed a steady flow into a series of scoops attached to a rotating wheel. The water's weight pulled the wheel forward in precise increments. This motion was transmitted upward through an ingenious chain drive, which Su Song invented centuries before the same principle appeared in Western mechanical clocks. The drive consisted of linked segments of metal that transferred power smoothly without slippage, a design that would not reappear in Europe until the 14th century.

Each motion of the chain advanced a complex gear system that controlled the celestial armillary sphere atop the tower. This sphere, open to the sky, mirrored the rotation of the heavens, allowing astronomers to observe the positions of the stars and planets in real time.

The Escapement: Time in Motion

At the center of Su Song's design was another revolutionary feature, the escapement mechanism, a device that regulated the release of energy to maintain constant speed. In Su Song's clock, a carefully balanced lever allowed the waterwheel to advance in steady, timed increments, preventing the motion from accelerating uncontrollably.

This was centuries before escapements appeared in European clockmaking. Modern historians now recognize Su Song's design as the earliest known example of mechanical regulation, a leap that transformed timekeeping from passive measurement into controlled precision.

The result was a machine that did not simply mark time, it embodied it. Each drip of water and each rotation of the gears represented a heartbeat of the cosmos.

The Display of Heaven

From dawn to dusk, Kaifeng's citizens could look up at the tower and see the passage of time reflected in motion. On the exterior, a towering face displayed rotating panels showing the hours, zodiac signs, and celestial phases. Bells rang automatically to mark the time, and wooden figures, beautifully carved and painted, emerged from hidden doors to strike gongs or present tablets.

These automata were not mere decorations; they were metaphors for cosmic order. Each appeared at precisely the right moment, as if summoned by the heavens themselves. To the people below, the tower was a living being, part temple, part observatory, part clockwork dragon whose breath was time.

Inside, astronomers recorded data from the armillary sphere and the star maps within. The precision of Su Song's measurements allowed the imperial court to reform the calendar, ensuring that official timekeeping matched celestial reality. In an empire that viewed cosmic alignment as the foundation of moral governance, the tower's accuracy was both scientific and spiritual.

Engineering Power and Harmony

The construction of the clock tower was a feat of national significance. Emperor Zhezong, who commissioned the project, viewed it as an emblem of the dynasty's sophistication. Its function went beyond astronomy, it was a demonstration of order in motion. Every perfectly timed bell and celestial rotation affirmed that the Song emperor ruled under Heaven's rhythm.

The tower's operation required collaboration among hundreds of artisans, metallurgists, and scholars. Bronze casters shaped precise gears with tolerances remarkable for their time. Carpenters constructed the tower's wooden frame to absorb vibration and maintain alignment. Mathematicians calculated the gear ratios needed to synchronize daily, lunar, and stellar cycles.

In this single project, the empire's intellectual and mechanical powers merged, a reflection of the Song dynasty's broader achievements in science, printing, and engineering.

Collapse and Rediscovery

For thirty five years, Su Song's masterpiece stood as the pride of Kaifeng. But in 1127, disaster struck. The Jurchen armies of the Jin dynasty invaded the north, capturing the city and seizing its treasures. The invaders dismantled the clock tower and transported its parts northward in an attempt to reconstruct it. Without Su Song or his skilled assistants, the machine could not be restored. The gears fell silent and the Tower of Heaven passed into legend.

Yet Su Song anticipated such loss. Before his death, he compiled a monumental text, Xin Yi Xiang Fa Yao (New Design for a Mechanized Armillary Sphere and Celestial Globe), containing meticulous descriptions and detailed diagrams of his clock. More than a thousand years later, these records remain among the most precise technical documents of the pre modern world.

Thanks to these diagrams, historians and engineers have reconstructed the tower on paper, in models, and even digitally. The 20th century mechanical engineer Joseph Needham, whose multivolume Science and Civilization in China reintroduced Su Song's work to the modern world, called it "one of the greatest monuments of mechanical engineering in history."

The Machine as Mandate

The symbolism of Su Song's clock tower reached far beyond its mechanics. To the scholars and courtiers of the Song dynasty, it represented the ideal harmony between human order and cosmic rhythm. In a society shaped by Confucian ethics and Daoist cosmology, time was not an abstraction, it was moral. The heavens were perfect, their cycles eternal; to emulate them was to align civilization with virtue.

The Tower of Heaven thus became a physical embodiment of the emperor's mandate. Its precision confirmed that Heaven smiled upon the dynasty. Each movement of the gears was both mechanical and metaphysical, a reminder that science and spirituality could coexist without contradiction.

The Echo of the Gears

Today, modern reconstructions of Su Song's clock tower stand in museums and research institutes, their bronze and wood replicas ticking softly in tribute to their creator. Visitors who watch the gears turn and the celestial sphere rotate experience a small measure of what Kaifeng's citizens once felt, awe not merely at invention, but at meaning.

The tower's mechanisms anticipated modern engineering principles by centuries, yet its purpose was deeply ancient, to express humanity's longing to understand its place in the universe. In Su Song's clockwork dragon, the cosmos itself seemed to breathe.

Through his creation, Su Song bridged earth and sky, knowledge and belief. Though the original tower was lost, its rhythm never truly stopped. It continues to turn in every clock, every observatory, every human effort to give structure to the infinite.

And so, in the stillness of history, if one listens closely, the faint echo of the gears remains, a steady heartbeat reminding us that even the heavens can be measured, if only for a moment, by the hands of man.

Chapter 6
Water and Wonder – Egypt's Hydraulic Innovations

Morning light shimmered across the slow current of the Nile. A group of engineers, their linen garments damp with river spray, worked in practiced rhythm. As one man pulled down the long wooden beam of a shaduf, another guided the clay bucket into the water. With a smooth counterweight motion, the beam lifted and the bucket poured its contents into a narrow channel. The flow wound its way through fields and courtyards, past workshops and into the walled precinct of a temple, where the same water that nourished the crops would soon breathe life into stone.

In ancient Egypt, water was not merely a resource. It was the pulse of existence, a divine element that linked the earthly to the eternal. It flooded and receded according to the rhythm of the gods. From this sacred source, Egyptian engineers drew both inspiration and power. Through their mastery of hydraulics, they tamed a river and, in doing so, gave motion to temples, fountains, and statues that seemed to move by their own will.

The Sacred River of Innovation

The Nile was Egypt's heartbeat. Each year, its floods renewed the soil and defined the calendar. But controlling its power required invention. By 3000 BCE, Egyptians had already learned to channel the river through irrigation basins and canals. Their tools, shadufs, water wheels, and sluices, were feats of balance and precision. These same techniques, developed for agriculture, soon found their way into spiritual spaces where they served purposes beyond survival.

For Egyptians, the natural and divine were inseparable. Water symbolized creation and renewal. The god Hapi, spirit of the Nile's flood, was depicted carrying offerings of grain and lotus flowers. Temples were built not only beside the river but in harmony with it. Hydraulic systems became expressions of worship, designed to animate the sacred with the same vitality that animated the land.

Water in the Service of the Gods

Within temple walls, engineers transformed irrigation techniques into marvels of ritual performance. They created self filling basins, automatic fountains, and statues that used the principles of pressure and balance to astonish worshippers.

One of the earliest hydraulic wonders was the water clock, or clepsydra. Used to measure time during religious ceremonies, these clocks functioned through a simple yet ingenious principle. A vessel, often shaped like a cone or decorated with divine imagery, allowed water to drip at a steady rate through a small hole in its base. As the level changed, markings on the interior indicated the passage of hours. Priests could track sacred intervals for prayers, offerings, and celestial observations.

Egyptian water clocks were refined over centuries. By the time of the New Kingdom, around 1400 BCE, the clepsydra had become both a scientific and artistic masterpiece. Some were inscribed with hieroglyphs representing the hours of the night, aligned with the rising of specific stars. In this way, time, astronomy, and devotion flowed together, measured drop by drop.

The Temple as Machine

Temples were the engines of Egyptian civilization, centers of economy, ceremony, and science. Their design often incorporated hidden channels and chambers that allowed water to perform miracles.

In the great temple complexes at Karnak and Edfu, archaeologists have uncovered evidence of subterranean conduits that connected courtyards, fountains, and sacred lakes. These systems allowed priests to fill basins or pour libations from hidden sources, creating the impression that the gods themselves had intervened.

Imagine standing in the hypostyle hall of a temple, the air heavy with incense and sunlight filtering through massive columns carved with lotus and papyrus. A priest approaches an altar and, as he chants, water begins to pour from a stone vessel though no hand touches it. To the worshippers, it is divine mystery. To the engineer priests behind the scenes, it is air pressure and gravity in perfect partnership.

Such devices were powered by the same principles later formalized by Greek engineers like Hero of Alexandria. But the Egyptians were first to harness those forces in the service of faith. They understood that wonder was not diminished by knowledge, it was magnified by it.

The Breath of the Gods

Among the most fascinating applications of Egyptian hydraulics were mechanical statues that seemed to move or speak. Accounts from later Greek travelers, such as Herodotus and Diodorus Siculus, describe temples where statues nodded or produced sounds during rituals. Some modern historians believe these effects were achieved through concealed tubes or chambers where heated air expanded and contracted, moving small parts or forcing sound through reeds, primitive pneumatics that predated Hero's temple mechanisms by centuries.

One particularly striking example may have been the Colossi of Memnon, two enormous seated statues near Thebes. In the early morning, as the rising sun warmed the stone, visitors reported hearing a haunting tone, like music or a human voice. The sound likely came from air escaping through cracks as the stone expanded, but to ancient ears, it was the god greeting the dawn. Even accidental hydraulics became evidence of divinity.

The Dual Nature of Water

Egypt's relationship with water was both practical and spiritual, scientific and symbolic. To irrigate a field was to imitate the act of creation itself. The annual flooding of the Nile was celebrated as the moment when chaos was tamed and order renewed. Hydraulic engineering thus carried deep religious meaning, it was a dialogue with the divine.

Priests who controlled temple hydraulics were more than technicians; they were custodians of sacred knowledge. Their understanding of flow and pressure was passed down through generations, often kept secret within temple schools. This dual role, scientist and priest, helped preserve Egypt's technological traditions long after their original purposes faded.

Hydraulic Artistry in Daily Life

Outside temple walls, hydraulic innovation shaped daily Egyptian life. Cities and estates used intricate systems of wells, canals, and cisterns to ensure a steady water supply. The shaduf, still seen along the Nile today, was both tool and symbol, its balanced beam a visible metaphor for Ma'at, the cosmic principle of order and equilibrium.

Public fountains and baths, though utilitarian, carried spiritual connotations of purity. Ritual washing before entering sacred spaces reflected the belief that water cleansed both body and soul. Even in the home, small household shrines sometimes featured miniature basins fed by trickling water, echoing the grand designs of the temples.

Engineering as Devotion

The Egyptians saw no boundary between science and religion. To them, engineering was a sacred act, a way of participating in the divine order. Every canal dug and basin filled echoed the first act of creation, when the primeval waters receded and the land emerged.

In this worldview, a machine was not a challenge to the gods, it was their servant. When water flowed through hidden channels to animate a statue or fill a basin, it reaffirmed the unity of physical law and spiritual truth. The temple's engineers were not mere craftsmen; they were theologians of motion, translating faith into form.

Echoes Through Time

The legacy of Egypt's hydraulic ingenuity reached far beyond its borders. Greek and Roman engineers later studied Egyptian water systems, adapting their principles to aqueducts, fountains, and mechanical entertainment. The continuity between Egyptian temple hydraulics and later automata reveals a lineage of curiosity that transcended empires.

Modern reconstructions of Egyptian water clocks and temple devices have confirmed their practicality. A replica of an 18th dynasty clepsydra, for instance, keeps remarkably accurate time when calibrated to the latitude of Thebes. The engineering precision required to carve these devices by hand, measuring flow rates, calculating vessel volume, and shaping uniform openings, speaks to a civilization deeply attuned to both measurement and meaning.

The Eternal Flow

As twilight settled along the Nile, the water shimmered like molten gold. The same river that had fed crops and carved valleys for millennia still whispered against its banks. The shadufs creaked in the distance, unchanged since the days of the pharaohs. In temple ruins, the faint outlines of basins and channels remain, silent testimony to the minds that once made stone come alive.

Through their mastery of hydraulics, the Egyptians achieved something greater than technology, they expressed faith through motion. To them, every drop of water carried a fragment of the divine spark, every flow a reminder that the boundary between earth and heaven was as fluid as the Nile itself.

In their world, water was not only life. It was wonder made visible.

Chapter 7
Minds of Metal – The Philosophy Behind Artificial Life

The afternoon sun slanted across an Athenian courtyard where a small group of men gathered under the shade of an olive tree. Scrolls lay open beside them, weighed down by smooth stones to keep the breeze from turning the pages. Voices rose and fell in debate. Could a machine that imitated motion, speech, or even thought truly possess a soul, or was it merely clever deception, a reflection of life without life itself? One man argued that anything capable of movement must have its own inner principle. Another countered that motion could be borrowed, transferred, or imitated through craft. The discussion was both philosophical and prophetic. These were not idle musings but the first recorded attempts to understand what, centuries later, would become the question of artificial intelligence.

The Soul of Motion

In ancient Greece, motion defined life. To move was to be alive, and to cause motion in others was to exert will. The earliest Greek philosophers wrestled with how this applied to the natural world. Thales, Anaximander, and Anaximenes, all among the Milesian thinkers of the sixth century BCE, saw in nature a living principle that animated all matter. They imagined the world itself as a kind of organism, pulsing with energy. If all things were infused with life, then a moving statue or self opening door might not violate natural law; it might reveal it.

By the time of Aristotle, the question of motion had become both mechanical and metaphysical. In his treatise On the Soul, Aristotle described life as the presence of a psyche, a soul that organized matter toward purpose. Plants had the lowest form of soul, governing growth and nutrition. Animals possessed sensation and movement. Humans alone combined reason with will. Yet Aristotle also observed that certain motions could occur without life. A falling stone or a spinning wheel was moved by causes external to itself. Motion alone, he warned, did not prove the presence of a soul.

Still, Aristotle was fascinated by the possibility of lifelike mechanisms. In his Politics, he imagined a world where "each instrument could accomplish its own work... the shuttle would weave and the plectrum play the lyre of itself." In such a world, he mused, masters would have no need of servants. He did not mean this as prophecy but as a thought experiment about causality and labor. Yet in envisioning self acting tools, Aristotle glimpsed what later centuries would call automation.

Plato and the Divine Craftsman

Plato approached the question of artificial life from another direction, through metaphor. In his Timaeus, he described the universe as a living being created by a divine craftsman, the Demiurge. This cosmic artisan shaped chaos into order and endowed the world with a soul. The heavens moved in perfect circles because they were alive, rational, and governed by mathematical harmony.

To Plato, imitation was not deception but participation in truth. When a sculptor carved a figure that seemed to breathe, he was not challenging the gods but reflecting their creative act. The maker of automata, likewise, revealed the hidden geometry of life. A machine that imitated motion did not possess a soul of its own, but it mirrored the rational order that defined existence. In this sense, the engineers of antiquity were philosophers in metal, their devices embodiments of Platonic form.

The Stoic Mechanism of Nature

Centuries later, the Stoics extended these ideas into a grander synthesis. They saw the universe as a vast, interconnected mechanism governed by logos, the divine reason or animating fire that permeated all matter. Every part of nature acted in perfect coherence, like the gears of a clock. The Stoics spoke of the cosmos as a self regulating system, an early conception of what we might call natural law.

Under this worldview, the distinction between living and nonliving blurred. If divine reason animated all things, then the motions of machines could be understood as extensions of that same cosmic force. A mechanical bird or self moving statue was not defying nature, it was imitating her. The builder who harnessed air, water, or heat to make matter move was, in a small way, participating in the divine logos.

This belief resonated with Hellenistic engineers such as Ctesibius, Philo of Byzantium, and Hero of Alexandria. Their machines were not merely practical, they were demonstrations of harmony between intellect and nature. When Hero wrote of air and water acting through pressure, he was not describing inert matter but living forces that obeyed predictable laws.

The Mechanistic Metaphor

Greek philosophers often used mechanical imagery to explain the workings of the body and mind. Plato compared the human soul to a charioteer guiding two horses, one noble and one unruly. Aristotle described the heart as a kind of furnace and the lungs as bellows that fed it. Later physicians in Alexandria likened the nervous system to a network of pipes carrying pneuma, vital air, through the body.

Such metaphors reveal how closely the Greeks associated life with mechanism. To them, the body was a device animated by divine air and the world itself was a body animated by cosmic breath. The mechanical and the spiritual were not opposites, they were reflections of the same mystery.

Engineers as Philosophers

While philosophers debated, engineers built. The craftsmen who created automata for temples and theaters were rarely named in surviving texts, yet their work embodied the very questions philosophers asked. Was a statue that poured wine truly alive? Was a door that opened in response to flame obedient to will or to nature?

For Hero of Alexandria and his predecessors, these were not abstractions. They designed systems that transformed invisible forces into motion, heat into pressure, water into power, and air into song. Each mechanism was an argument made in metal, a demonstration that reason could command the physical world.

It is no coincidence that Hero's era coincided with the flourishing of Stoic philosophy. Both celebrated the predictability of nature. Both believed that understanding the laws of motion brought one closer to divinity. Hero's devices, described in his Pneumatica, are filled with the same reverent logic that animates Stoic cosmology. The world breathes, therefore it moves. It moves, therefore it lives.

The Ethical Edge

But even in antiquity, the imitation of life raised moral questions. Some thinkers warned that to create lifelike machines was to trespass upon divine prerogative. Others saw in such creation the highest expression of human intellect. The myth of Prometheus, who stole fire from the gods to give life to clay, symbolized both triumph and punishment. He gave humanity knowledge and suffered eternally for it.

In this mythic frame, every automaton carried a hint of defiance. A machine that moved by itself blurred boundaries between art and life, and between creator and creation. When artisans built statues that nodded, bowed, or poured libations, they reenacted Prometheus's act on a smaller, safer scale. Yet the same tension lingered. How far could human ingenuity go before it rivaled the divine?

The Living Cosmos

Underlying all these debates was a shared conviction that the universe itself was alive. From the orderly motions of the stars to the circulation of blood in the body, the Greeks saw life as pattern. Machines that imitated those patterns were not profane, they were reverent. Each mechanism was a microcosm, a model of the greater whole.

When an automaton turned its head or sang through hidden pipes, it echoed the celestial spheres turning in their orbits. To understand motion was to understand the mind of the cosmos. In that sense, the philosopher and the engineer were partners: one sought truth in words, the other in wheels.

The Echo Through Time

These philosophical foundations endured long after the physical machines rusted away. Medieval scholars, reading Aristotle through Arabic translations, continued to discuss the relationship between soul and motion. Renaissance thinkers revived Plato's Demiurge and the Stoic logos, interpreting them through mechanical analogies. By the seventeenth century, when Descartes described animals as living machines and the universe as a clock, he extended an idea that had begun two thousand years earlier in an Athenian courtyard.

The Greeks never built a thinking machine, but they built the idea of one. Their myths, metaphors, and mechanisms all pointed toward a single question that still lingers in our age of robotics and algorithms. When does imitation become creation?

The Soul in the Gears

As the debate in the courtyard drew to a close, the philosophers looked toward the fading light. The olive leaves trembled in a faint wind, their motion gentle and ceaseless. One of them smiled and gestured toward the tree. "Perhaps the question is not whether a machine can have a soul," he said, "but whether the soul itself is a kind of machine."

In that moment, the line between thought and motion, between life and mechanism, dissolved into the golden haze of the evening. The Greeks had not built robots, but they had built the framework for understanding them. In their world, to think was to move, and to move was, in some mysterious way, to live.

Chapter 8
The Secret of the Scribes – Alexandria's Hidden Workshops

A narrow staircase wound downward beneath the great Library of Alexandria, away from the noise of scholars and the scent of papyrus. A lone scribe carried a lamp, its flame flickering against walls carved with ancient equations and notes in Greek, Egyptian, and Aramaic. At the bottom, the passage opened into a small chamber. On tables of cedar and bronze lay strange instruments, lenses of polished crystal, fragments of tubing, cogs, weights, and vessels half filled with mercury. The scribe paused, awed. This was no ordinary archive. This was the workshop of the library itself, the hidden heart of the world's knowledge, where theory became invention.

The City of Reason

In the third century BCE, Alexandria stood as the beating heart of human intellect. Founded by Alexander the Great and perfected under the Ptolemies, it was a city built to embody the Greek pursuit of knowledge and the Egyptian love of permanence. Its streets teemed with scholars, mathematicians, physicians, astronomers, and artisans. Ships in its harbor brought not only spices and papyrus but also manuscripts and ideas from every corner of the known world.

At its center rose the Library and the adjacent Mouseion, or Temple of the Muses, a sprawling complex of lecture halls, gardens, observatories, and laboratories. It was more than a library. It was a university centuries before the term existed, a living engine of discovery. Within its walls, human curiosity became organized knowledge.

Here worked Euclid, who defined the geometry that still shapes architecture and engineering. Eratosthenes measured the circumference of the Earth with astonishing accuracy. Archimedes, though based in Syracuse, sent his works here to be copied and studied. The Mouseion gathered the world's greatest minds, and among their pursuits was a growing fascination with mechanics, the study of motion, pressure, and the unseen forces that governed them.

The Mechanics of the Mind

In the workshops beneath the colonnades, inventors and philosophers collaborated freely. The separation between science and craft did not exist. To design a water clock was as honorable as to compose a theorem. The scholars of Alexandria understood that abstract knowledge gained meaning only when embodied in mechanism.

One of the earliest mechanical pioneers associated with the library was Ctesibius, an engineer and mathematician of the third century BCE. The son of a barber, he reportedly discovered the principle of air pressure while repairing a mirror stand in his father's shop. His experiments with compressed air led to inventions such as the hydraulic organ, the force pump, and the self regulating water clock, devices that transformed both entertainment and daily life.

Ctesibius's students included Philo of Byzantium, whose Mechanics and Pneumatica described early versions of pressure powered automata and water driven toys. Philo's treatises, preserved partly through later Arabic translations, formed the bridge between the inventors of classical Greece and the later genius of Hero of Alexandria.

The Library as Laboratory

The Library of Alexandria was not merely a repository for scrolls but a crucible for experimentation. The scholars there pursued not only mathematics and astronomy but also the practical sciences of optics, hydraulics, and pneumatics. They experimented with light, mirrors, and lenses, studying reflection and refraction. It is believed that early prototypes of burning mirrors, giant concave lenses said to focus sunlight into beams hot enough to ignite ships, were first designed in Alexandria.

The Mouseion's workshops may also have contained rudimentary devices for measuring air and water flow, balance mechanisms, and automata used for demonstration. These would have served not as entertainment but as proof of theory, physical manifestations of ideas discussed in the lecture halls above.

There are even fragmentary accounts suggesting that certain sections of the library were dedicated to what we would call engineering archives, collections of models and drawings used for teaching. If so, then the loss of the Library was not only the loss of words but also of objects, tools, and prototypes that embodied the earliest mechanical imagination.

The Web of Knowledge

The exchange of ideas within Alexandria's walls was astonishingly fluid. Greek mathematicians corresponded with Egyptian priests versed in practical engineering, who in turn exchanged insights with Babylonian astronomers and Persian metallurgists. Translators rendered texts from Hebrew, Sanskrit, and hieroglyphic sources into Greek, merging empirical observation with theoretical reasoning.

This fusion gave rise to the Hellenistic scientific method, in which philosophy, mathematics, and experiment were interwoven. Mechanics, optics, and pneumatics became part of a single continuum of inquiry, the study of how invisible forces shaped visible phenomena. The Alexandrian scholars did not see the material world as static. They saw it as a living system of cause and effect, one that could be understood, replicated, and even improved.

Within this worldview, the mechanical devices described by Ctesibius and Philo were not novelties. They were demonstrations of universal law. A self filling basin or a singing bird revealed as much about the cosmos as a telescope or a theorem.

The Shadow of Secrecy

Despite the library's commitment to knowledge, much of its mechanical work was kept hidden from public view. Many of the automata and hydraulic devices were designed for temples, where secrecy was essential to sustaining the aura of divine power. The engineers who devised these wonders likely operated in a delicate balance between science and sanctity.

Some historians speculate that special chambers within or beneath the Mouseion may have been reserved for technical experiments—safe spaces away from prying eyes and combustible scrolls. If such workshops existed, they would have contained models of water pumps, early vacuum experiments, and perhaps even primitive lenses used for magnification or focusing light. The fire that ultimately consumed the Library erased not only the written record but also the very instruments through which the ancients explored the world.

The Fragile Record

The fate of the Library of Alexandria remains one of history's great tragedies. Whether it perished in a single catastrophic blaze or in a series of gradual losses, the result was the same: the silence of thousands of voices. The works of Ctesibius survive only through quotations. Those of Philo are fragmentary. Countless others have vanished altogether.

Among the lost scrolls were likely treatises on optics, mechanical motion, and hydraulics, texts that could have advanced human knowledge by centuries. One can only imagine what devices might have been described there, improved clockwork mechanisms, designs for automatic irrigation systems, or theories on the transformation of heat into movement.

The Library's destruction was not just the burning of books. It was the extinguishing of experiments still in progress. The hidden workshops where scribes, engineers, and philosophers once collaborated fell silent. In the ashes of papyrus and bronze, the line between myth and science blurred once again.

Echoes of Genius

Yet fragments of that lost brilliance survived. Copies of Alexandrian works carried to Pergamon, Byzantium, and later to the Islamic world preserved the core of Hellenistic mechanical science. When scholars in Baghdad's House of Wisdom translated Greek texts into Arabic, the legacy of Alexandria lived anew. Hero's Pneumatica, the culmination of centuries of mechanical study, became a beacon for engineers from Damascus to Al-Jazari's court.

The spirit of the Library, the union of curiosity and craft, never truly vanished. Its principles reemerged in medieval water clocks, Renaissance optics, and the first telescopes that again sought to measure the heavens. Each innovation whispered of the rooms once hidden beneath the colonnades of the Mouseion, where scribes and inventors turned thought into movement and theory into flame.

The Flame That Endures

If one could walk again through the ruins of the ancient Library, the air might still smell faintly of oil and ink. In imagination, the stairway to the subterranean workshop remains, its stone steps worn by generations of scholars. Somewhere below, a water clock drips in darkness, its rhythm unchanged for two thousand years.

The Library of Alexandria may have fallen to fire, but its secret workshops never truly died. They live on in every laboratory and observatory where curiosity is given form, in every gear that turns, every beam that refracts, every question that refuses to burn.

For in the end, the greatest mechanism ever built in Alexandria was not of bronze or glass. It was the human mind itself, restless, intricate, and forever reaching toward the light.

Chapter 9
The Emperor's Clockwork Court — Mechanized Marvels in Imperial China

Silk rustled like water as courtiers took their places along the polished floor. Lantern light floated across lacquered screens, catching the gold of inlaid dragons. Musicians raised flutes. Then, as if answering an invisible cue, a fountain arced in a perfect silver curve and a pair of jeweled birds fluttered from a carved bough to sing in alternating trills. A lotus near the dais slowly opened, petal by petal, to reveal a tiny dancer who began to turn with the measured grace of a star. No stagehands crossed the hall. No servants hurried forward with ropes or poles. The spectacle unfolded by its own hidden pulse, the palace itself a living instrument that kept time with the emperor's breath.

From the Han to the Tang dynasties, imperial China cultivated a tradition of mechanical entertainment that blended court ceremony, natural philosophy, and the politics of wonder. The devices that turned and sang at banquets were not idle toys. They were proofs in motion, demonstrations that technique could harmonize with cosmology, that the sagely ruler presided over a world whose rhythms were orderly and precise. The entertainments delighted guests, but their deeper purpose was to make legitimacy visible.

Ancestors of the Art: A Humanoid in a Bronze Age Story

Chinese writing preserves one of the world's oldest tales of an artificial being. In the anthology later known as the Liezi, a craftsman named Yan Shi appears before King Mu of Zhou with a life sized automaton. The figure walks, bows, sings, and even flirts, until the king, unnerved by its apparent consciousness, orders it opened. Inside the skin the court finds wood, leather, and lacquer, each part labeled for function: eyes for sight, throat for song, intestines for breath. The king praises the artificer and spares the device.

The story is literature, not engineering report, compiled centuries after the period it depicts. Yet it is revealing. It shows that by early classical times Chinese thinkers could imagine a body as an assembly of discrete mechanisms, each with a role in a coordinated whole. The tale's vocabulary of hoses, valves, and articulated joints anticipates later craft descriptions. More importantly, it ties mechanical ingenuity to kingship. An automaton in the throne room is not merely a marvel; it is a mirror of rule, a model of the body politic obedient to a single will.

Music in Water and Air: Han Foundations

Under the Western Han (206 BCE to 9 CE), court spectacle matured into an art of precision. Hydraulic devices timed processions and framed ritual. Water clocks, clepsydrae with float indicators and gear trains, regulated guards and ceremonies. Court mathematicians developed calendars that set hours by the flow of water, and artisans learned to stabilize that flow with compensating tanks and drip regulators. When a bell rang or a drum sounded without a visible hand, it did so because the palace had become a measured environment.

Entertainments followed the same logic. Accounts from later histories describe singing birds and moving animals presented at feasts. Whether powered by concealed bellows or by siphons that forced air through whistles, the effect was the same: nature reproduced under human control. The palace stage extended into gardens whose pavilions hid pipework beneath paving stones. Jets leapt at a clap or a cue from the orchestra, and water turned thin metal leaves that flashed like fish. The audience beheld a domesticated cosmos, breath, flow, and rhythm harmonized under order.

A Court Engineer of Genius: Ma Jun and the Theater of Motion

During the Three Kingdoms period, when emperors and warlords competed for cultural as well as military prestige, the engineer Ma Jun, active in the third century CE, became emblematic of mechanized spectacle at court. His reputation rests on multiple achievements: an improved chain pump for irrigation, a refined south pointing chariot that kept a figure oriented despite turns, and, in the context of entertainment, mechanical theaters and displays that astonished the Wei court. Later technical treatises describe a water driven puppet stage in which miniature figures rowed boats, beat drums, or raised banners while mountains emitted sound.

If the descriptions read like fantasy, their mechanisms do not. Ma Jun had access to pumps, siphons, waterwheels, and wind chests. A pressure head established by a tank above the hall could drive multiple motions in parallel. Floats could trigger cams and levers at predictable levels. Air, forced through flutes by head pressure, could give each puppet a voice. The show's power source was neither divine nor mysterious. It was the same river logic that irrigated the empire. In Ma Jun's hands, hydraulics became a language of courtly celebration.

Timed Wonder as Governance

Rulers understood the political charge of a machine that never erred. When an emperor's banquet unfolded under the governance of a clock regulated stage, he appeared to preside over inevitability. Every figure entered at the right instant, every fountain rose at the precise stroke. The guests experienced abundance, but also discipline. The spectacle expressed a conviction drawn from the Confucian classics and elaborated by court astronomers: that moral rule aligned human affairs with Heaven's cycles.

This conviction bound entertainment to measurement. A palace display drew its power from the same timing systems that kept the calendar true. The Bureau of Astronomy provided ratios for gear trains and periods for celestial indications. Artisans translated those periods into cams, star wheels, and peg drums. To synchronize music, motion, and light was to show that the ruler kept time with the sky.

Tang Splendor: Motion in a World Capital

By the Tang dynasty (618 to 907), Chang'an became the most cosmopolitan city in the world. Caravans arrived with glass and silver. Envoys brought musicians and magicians. Buddhist monks carried automaton lore from temple to temple. The court of Emperor Xuanzong turned performance into policy, founding the Pear Garden academy and filling palace grounds with pavilions where sound, color, and movement confirmed imperial radiance.

Mechanical entertainments multiplied. Descriptions speak of lotus standards whose petals opened to reveal lanterns, mounts and beasts that moved with hidden joints, fountains that kept even beat with instruments. Within this culture of spectacle, Yi Xing, a monk astronomer, and Liang Lingzan, an imperial artisan, designed a water regulated armillary sphere with a mechanical escapement, a device that released power in measured increments and kept the heavens' model turning at the correct rate. Though dedicated to observation rather than entertainment, their instrument belonged to the same aesthetic of timed motion that governed the banquet hall. A palace that could measure a star could time a dance.

The Tang also refined incense clocks, long branching trails of aromatic powder that burned at calculated rates, dropping small balls onto gongs or releasing threads tied to miniatures that moved when a string parted. In temple liturgies and night watches alike, the gentle fall of a bead or the glide of a figure marked time without water's sound. Such devices easily crossed from monastery to palace, where a drifting fragrance and a quiet chime could regulate an hour in the midst of revelry.

Devices in the Texts: Where Story Meets Craft

Chinese sources do not always separate demonstration from delight. Technical books catalog the workings of pumps, valves, and wind chests while anecdotal histories recount mechanical wonders at court. The Yan Shi automaton remains a philosophical parable. Mozi and Lu Ban appear in other texts credited with wooden birds that flew for days, almost certainly impossible, yet suggestive of gliding kites and the habit of attributing marvels to culture heroes. By contrast, the force pump, the hydraulic organ, the geared sphere, and the puppet theater lie within the range of materials and methods documented in surviving treatises and later reconstructions.

The line between myth and mechanism was not a weakness of the tradition. It was the tradition's engine. Stories invited craftsmen forward, proposing effects that texts then explained in the sober language of head, flow, and leverage. Court art existed in a feedback loop with natural philosophy: an emperor asked for a vision, a scholar calculated the intervals, an artisan built the gears, a historian described the result, a philosopher gave it meaning.

Techniques Behind the Curtain

The mechanics that animated palace entertainments were direct descendants of agricultural and timekeeping arts. A typical banquet display might begin in a cistern above the hall whose outlet fed a header tank. From there, water ran through hidden lead or bamboo pipes to wind chests that forced air through whistles. A float riding in a small chamber rose at a known rate, turning a peg drum that advanced cams. Each cam lifted a lever. Each lever raised an arm, opened a beak, or released a valve to start a sparkling jet.

Where water flow needed smoothing, craftsmen added compensators, secondary tanks and pinch points that damped surges. Where motion risked stutter, they interposed regulators or escapements derived from astronomical instruments. Materials mattered. Lacquered wood resisted swelling and warp. Bronze held fine teeth. Silk cord carried loads with minimal stretch. Oiled leather made hinges silent. Nothing in the apparatus was accidental. The devices borrowed the entire material vocabulary of the empire and rearranged it around a single purpose: to make time tangible.

Meaning for the Throne

Spectacle's philosophy was simple. When clockwork birds alternated their song with regularity, they enacted reciprocity under the ruler's gaze. When a fountain's jets rose in rank and sank without spill, they modeled hierarchy. When a lotus opened at the moment the master of ceremonies named an auspicious hour, it signaled harmony between decree and fate.

In Confucian terms, such performances illustrated li, the pattern of ritual propriety, each act in its place, each sound at its time. In Daoist terms, they revealed ziran, the self-so of things, water flowing because it flows, air singing because it must. The court gathered both readings into a single message. The ruler who could host a universe of coordinated motions was himself in step with the Mandate of Heaven.

Preservation, Loss, and Reconstruction

Unlike the massive astronomical clocks of later Song times, the entertainments of Han and Tang courts were often temporary, assembled for a feast, repurposed for a festival, then reused or lost. War, fire, and decay erased most evidence. What survives are descriptions in histories, technical notes embedded in broader treatises, and the logic of materials that later engineers, East and West, have used to rebuild the effects. Modern reconstructions of float controlled fountains, incense clocks, and water driven puppet drums confirm how little is required to produce the coordinated displays described in the chronicles. The principle is modest. The impression is immense.

Loss also transformed the tradition into allegory. When later writers remember a lotus that opened by itself or a bird that sang at midnight, they record an engineering practice already receding from common experience. What had once been craft, passed from master to apprentice, became memory framed as miracle. Yet the memory keeps a core of truth. The entertainments were real enough to move dignitaries and to secure the reputations of the artisans who made them.

The Hall that Breathed

The palace performance returned again and again to a single theme: life in the inanimate. A hall that breathed with fountains and whistles was more than decoration. It was an argument that the world itself, rivers, winds, stars, could be understood, measured, and joined to human ceremony without violence to its nature. In such a hall, the emperor dined inside a model of the cosmos, and the guests, hearing birds that were metal and seeing flowers that were wood, learned that artifice and reality were partners rather than rivals.

So the jeweled birds resumed their alternating song, the lotus closed and opened on cue, and the fountains rose and fell as the musicians marked the beat. Somewhere behind a screen a float lifted a lever, a cam released a pin, and a jet found its shape in air. The court watched, enraptured, as motion answered order and order returned as joy. The devices disappeared into their effects, and the effects resolved into meaning: a ruler in tune with the clockwork of Heaven, a city that could make water dance, and an empire that knew how to turn knowledge into wonder.

Chapter 10
The House of Knowledge – Islamic Engineers and the Revival of Mechanics

In a sunlit hall of Baghdad's Bayt al Hikma, the House of Wisdom, a scribe sharpened his reed pen and leaned over a scroll. Across the table, another scholar traced a diagram of interlocking wheels, their teeth marked with precise ratios. The sound of flowing water echoed from a courtyard fountain, its steady rhythm matching the murmur of scholars debating mathematics, astronomy, and the science of movement. This was no ordinary library. It was the engine of a new mechanical age. Here, the lost manuscripts of Greece found new life, and from them emerged the greatest inventors of the medieval world.

A Civilization of Curiosity

By the ninth century, Baghdad had become the intellectual heart of the Islamic world. Under the Abbasid Caliphs, especially al Ma'mun, knowledge became both spiritual and political pursuit. The House of Wisdom gathered scholars from Persia, Syria, Egypt, and beyond, who translated and expanded upon Greek, Persian, and Indian works. The writings of Hero of Alexandria, Ctesibius, and Archimedes, texts once housed in Alexandria's lost library, arrived in Arabic, annotated, corrected, and transformed.

But the scholars of Baghdad were not content merely to preserve the past. They built upon it. Their experiments joined Greek geometry with Persian craftsmanship, producing fountains that sang, clocks that told stories, and machines that moved with precision.

The Banu Musa Brothers: Masters of Mechanical Marvels

Among the first to transform theory into practice were the Banu Musa brothers, Muhammad, Ahmad, and al Hasan, three polymath sons of a court astronomer. Working under Caliph al Ma'mun in the ninth century, they directed teams of artisans and engineers in the creation of automated instruments that astonished visitors.

Their landmark work, The Book of Ingenious Devices, described more than one hundred mechanisms, many derived from Greek sources yet distinctly improved. Within its pages were self trimming lamps, automatic fountains, and flute playing automata powered by air and water pressure.

One famous design, a mechanical servant who poured wine or water from a pitcher, used a hidden float valve to control flow. When the vessel tilted, air entered a chamber, forcing liquid through a spout until the float rose high enough to seal the passage. When the servant returned the jug to its resting position, the float dropped, resetting the system. To the casual observer, the machine appeared alive, responsive, and polite.

The Banu Musa also devised programmable fountains that alternated jets, changed shapes, or produced rhythmic bursts of water. Some even featured whistling birds, echoing Hero's designs but with improved pressure control that allowed greater range and duration. These creations demonstrated deep understanding of hydraulics, pneumatics, and feedback systems, concepts that would not reappear in Europe until the Renaissance.

The House of Wisdom as Workshop

The House of Wisdom was both an academy and an engineering workshop. Scholars debated philosophy upstairs while artisans built downstairs. The Banu Musa brothers employed mechanics who tested materials, refined valves, and calibrated flow rates in pools that gleamed under lamps of polished copper. Each invention began as a drawing, sketched in ink and annotated with mathematical proportions.

The intellectual atmosphere encouraged experimentation. Greek thought provided the foundation, but Islamic scholars merged it with practical observation drawn from irrigation systems, astronomical instruments, and musical theory. The result was an integrated vision of science, one that united motion, sound, and water in a single aesthetic of harmony.

Al Jazari: The Engineer of Elegance

Three centuries later, this tradition reached its height with Ismail al Jazari, chief engineer to the Artuqid court in Diyarbakir, in modern Turkey. Around 1206, he completed his masterwork, The Book of Knowledge of Ingenious Mechanical Devices, a richly illustrated manual of automata and timekeeping machines.

Al Jazari's creations transformed utility into art. His elephant clock, for example, was a towering symbol of cosmopolitan synthesis. The elephant represented India, the phoenix Persia, the waterwork Greece, and the turbaned automaton Islam. Inside, a water timed mechanism caused a ball to drop every thirty minutes, triggering a cascade of movements. A scribe raised his pen, a bird chirped, a mahout struck a cymbal. The spectacle embodied the unity of knowledge across cultures.

Even more remarkable were his programmable devices, mechanical musicians and servants whose actions could be altered by rearranging pegs and cams, the medieval equivalent of coding. His water powered peacock fountain dispensed soap and towels to guests through a sequence of automatic motions. Each step depended on the previous one, forming a chain of logic that modern engineers recognize as sequential control, the foundation of robotics and computation.

The Science of Time and Flow

The water clock, perfected over centuries of experimentation, became the symbol of Islamic mechanical genius. From Damascus to Cordoba, engineers created monumental timepieces powered by controlled flow and intricate gearing. Al Jazari's clocks, along with those described by his predecessors such as al Muradi and Ibn al Shatir, displayed not only hours but also zodiac signs, moon phases, and planetary positions.

In these devices, flow regulation reached new precision. Adjustable sluices allowed constant pressure, while gear ratios translated linear motion into rotation. The escapement mechanism, seen earlier in Su Song's tower, appeared here in evolved form. The clock's surface often was animated by automata, falcons dropping balls, doors opening to reveal musicians, or astrolabes turning in rhythm with celestial cycles.

Each motion had both practical and philosophical significance. To measure time was to measure creation. The continuous yet controlled flow of water symbolized the order of divine law, unceasing and exact.

From Baghdad to Europe

Through translation and trade, the mechanical arts of the Islamic world spread westward. In the twelfth and thirteenth centuries, scholars in Toledo, Palermo, and Constantinople translated Arabic texts into Latin, introducing Europe to the works of Hero, the Banu Musa, and al Jazari.

These manuscripts reached figures such as Villard de Honnecourt, whose thirteenth century notebook includes sketches of automata resembling Arabic models. By the Renaissance, engineers like Giovanni de Dondi, Taccola, and Leonardo da Vinci built on the same principles, float valves, cams, and chain drives, that had flourished in Baghdad centuries earlier.

Leonardo's notebooks, filled with musical fountains and mechanical lions, echo the world of al Jazari. His self propelled cart and automatic drummer may never have existed without the diffusion of Islamic mechanical knowledge.

Beyond Invention: The Meaning of Motion

In Islamic philosophy, invention was an act of understanding creation itself. Scholars such as al Kindi and al Farabi saw geometry, proportion, and motion as expressions of divine order. Machines that moved with grace were not attempts to rival God but to reflect His perfection through reason and craft.

For this reason, mechanical devices often adorned mosques and gardens as symbols of balance. A fountain that alternated its jets demonstrated unity within multiplicity. A clock that showed the phases of the moon connected daily prayer to the rhythm of the cosmos. In every sense, the Islamic automaton was both science and devotion, calculation and wonder intertwined.

The Legacy of the House of Wisdom

When Mongol armies sacked Baghdad in 1258, the House of Wisdom was destroyed. Many of its manuscripts were lost to the Tigris, their ink said to have darkened the river for days. Yet the knowledge did not vanish. Copies and translations had already spread across the Islamic world, from Cairo to Cordoba, and into Europe.

The legacy of those scholars, their water clocks, fountains, and automata, became the mechanical foundation of the Renaissance. The precision engineering of cathedral clocks, the hydraulic gardens of Italy, even the concept of mechanical programming all trace their ancestry to the Islamic golden age of mechanics.

The Continuum of Genius

As twilight falls over modern Baghdad, few stones remain of the old House of Wisdom. Yet its spirit endures in every laboratory, observatory, and workshop where curiosity meets craft. The same logic that guided al Jazari's fountains governs the circuits of today's robots. The same wonder that inspired the Banu Musa still drives inventors who dream of motion with purpose.

In the quiet rhythm of gears and water, the scholars of Baghdad discovered something timeless, that the universe itself is a machine of elegance, and the greatest act of faith is to understand how it moves.

Chapter 11
The Caliph's Elephant – The Gift That Measured Time

The elephant stood at the center of a golden hall, its polished bronze skin gleaming beneath hanging lamps. The murmuring crowd of courtiers fell silent as a scribe lifted his hand toward the water basin balanced atop the creature's back. A silver ball dropped, struck a cymbal, and set in motion a spectacle of synchronized precision. A falcon dipped its beak, a rider bowed, and a hidden automaton inside the tower released another ball to begin the next cycle. The elephant clock of Al Jazari measured not only the passing of minutes but the harmony of civilizations.

A Machine of Many Worlds

When Al Jazari unveiled his elephant clock in the early thirteenth century, he was offering more than a marvel of engineering. He was presenting a symbol of unity in an age of diversity. Each part of the design carried cultural meaning. The elephant itself, an emblem of wisdom and strength, evoked India. The tower that rose from its back was shaped in the style of a Persian pavilion. At its peak perched a phoenix, representing the spirit of renewal known from Persian mythology. The waterwork inside the tower drew inspiration from Greek hydraulic principles described by Hero of Alexandria, while the figure seated atop the elephant wore the turban and robes of an Islamic scholar.

In a single mechanism, Al Jazari blended the heritage of four civilizations, Indian, Persian, Greek, and Arab, into a living metaphor for knowledge as shared inheritance. It was an argument in bronze for the interconnectedness of the medieval world, expressed not in words but in motion.

The Engineer and His Masterpiece

Ismail Al Jazari, court engineer to the Artuqid rulers of Diyarbakir in present day Turkey, wrote his Book of Knowledge of Ingenious Mechanical Devices in 1206. The manuscript, richly illustrated with diagrams and color washes, described fifty mechanical designs ranging from fountains to water clocks and automata. Yet the elephant clock stood above them all as his crowning achievement.

It combined hydraulic timing, mechanical control, and theatrical display with precision unmatched in its era. Modern reconstructions have confirmed its accuracy. The clock could mark roughly half hour intervals with steady regularity using only water pressure, gravity, and simple gearing.

How the Clock Worked

At the heart of the elephant clock was a water powered timing system. Hidden inside the elephant's body, a sealed vessel slowly filled through a narrow hole. Inside floated a bowl with a small opening in its base. As water seeped in, the bowl sank at a steady rate, pulling a string connected to a series of levers and pulleys. When the bowl reached the bottom, it triggered a release mechanism that set off the visible spectacle above.

A ball dropped from the tower, rolling down a ramp into the mouth of a waiting falcon. The falcon tipped forward, dropping the ball onto a drum that sounded the passage of time. At the same moment, the elephant's mahout struck his cymbal. Doors opened in the tower, revealing miniature figures that rotated in sequence. The scribe raised his pen as if recording the moment, and the process began again when the vessel inside the elephant was reset.

To maintain continuous operation, attendants refilled the reservoir and re floated the bowl at regular intervals. The balance of hydrostatics and mechanical feedback ensured a reliable rhythm, a masterpiece of analog automation long before the emergence of the escapement clock in Europe.

The Philosophy of Harmony

Al Jazari understood symbolism as well as science. The elephant clock was more than a demonstration of craft. It was a philosophical statement about balance and interdependence. Just as water flowed through channels to move each part in turn, so knowledge flowed between civilizations, animating all. The machine's steady cycles reflected the order of the cosmos as described by Islamic astronomers, motion without chaos, variety without conflict.

The use of water to regulate motion carried theological meaning. In Islamic thought, water represented life, purity, and divine order. To harness it with such precision was to reflect God's mastery of creation. To Al Jazari, the regularity of his clock mirrored the reliability of divine law.

The Art of Multicultural Design

Every detail of the clock expressed unity through diversity. The Indian elephant formed the foundation, its strength supporting the structure above. The Persian palace tower symbolized imperial refinement and architectural beauty. The Greek hydraulic system embodied reason and scientific inquiry, while the Arab astronomer and scribe represented wisdom and record keeping.

Even the materials carried meaning. Brass and bronze from Middle Eastern traditions blended with Indian ivory and Persian enamel in some versions. Water flowed through channels shaped like lotus stems and phoenix feathers. The result was not a mixture of cultures but a balanced harmony, an early vision of global collaboration.

Engineering Precision

Modern engineers who have reconstructed the elephant clock, such as those at the Ibn Battuta Mall in Dubai and the Museum of Islamic Art in Doha, have confirmed Al Jazari's skill. The system's timing depends on predictable water flow, achieved through a calibrated orifice and constant head pressure. The descending bowl's motion translated through pulleys to lift small counterweights, releasing energy in measured steps. This concept anticipates the later development of mechanical escapements that would guide European horology.

The clock's programmable sequence of actions, ball drop, cymbal strike, bird motion, and door opening, demonstrates early understanding of logical automation. Each action triggered the next in a chain of cause and effect. Replace water with electricity and levers with switches, and the elephant clock becomes a primitive form of automated control.

The Gift of Time

Historical accounts suggest that Al Jazari designed the elephant clock not only for the Artuqid court but as a diplomatic gift for a caliph. In medieval Islamic society, clocks were more than instruments. They were political statements. A ruler who commanded time showed mastery over nature and civilization. The elephant clock, with its blend of artistry and mathematics, symbolized not domination but harmony, rule aligned with cosmic order.

When the device performed, courtiers witnessed more than entertainment. They saw proof that faith, science, and beauty could coexist. The clock did not proclaim authority loudly. It expressed it through rhythm.

Influence Beyond the Islamic World

The reach of Al Jazari's work extended far beyond Diyarbakir. His manuscripts were copied and translated across the Islamic world, from Syria to Egypt and Andalusia, and fragments eventually reached Europe through Arabic Latin translations in Spain and Sicily.

The principles behind the elephant clock inspired later European engineers, including Giovanni de Dondi, whose fourteenth century astronomical clock applied similar gearing and cyclic motion. Renaissance automata, mechanical theaters, animated fountains, and music boxes trace their lineage to the same hydraulic logic. Even Leonardo da Vinci's sketches of self moving devices carry the conceptual imprint of Al Jazari's sequential design.

A Clock as a Worldview

The elephant clock is remembered not only for its technical sophistication but also for its message. At a time when much of Europe struggled to rebuild after centuries of turmoil, the Islamic world had become the steward of global knowledge. The House of Wisdom in Baghdad, the observatories of Samarkand,

and the workshops of Damascus all contributed to a shared mechanical tradition.

The elephant clock stood as its emblem, a machine that translated intellectual ideals into movement. Every half hour, as the ball dropped and the figures stirred, it reminded onlookers that the world's wisdom was interconnected. Greek geometry, Indian numerals, Persian artistry, and Arab craftsmanship worked together to measure time.

The Rhythm That Endures

Centuries later, replicas of the elephant clock continue to draw crowds. Visitors watch the same synchronized dance of levers, weights, and flowing water that once captivated a caliph's court. They witness more than a reconstruction. They see the heartbeat of a civilization that understood time as both a mechanical and spiritual order.

In Al Jazari's creation, the universe found mechanical expression. The elephant's steady rhythm echoed the cosmic cycles charted by astronomers and the prayers recited by believers. The spectacle of motion became an act of reflection, a reminder that progress is not a single culture's invention but the shared legacy of humanity.

When the final ball rolled and struck the cymbal, the hall filled with the soft resonance of bronze. For a moment, all eyes lifted, not to the elephant nor the scribe, but to the rhythm itself. Time, once an abstract flow, had been made visible. In that measured sound, Al Jazari's genius still speaks, a whisper from the thirteenth century telling the modern world that knowledge, like water, finds its balance when it flows freely among all who seek it.

Chapter 12
The Roman Stage – Machines of Empire

The roar of the crowd filled the Colosseum like thunder rolling across stone. From the emperor's box, a signal was given, and beneath the sanded arena floor, a series of gears, pulleys, and counterweights shuddered to life. With a groan of timber and a hiss of concealed hydraulics, a trapdoor opened. Out of darkness rose a gilded platform bearing a cage of lions, their growls echoing through the vaults below. Moments later, an artificial forest with mechanical trees unfolded from hidden slots, transforming the arena into a living landscape. Every shift of scenery and every appearance of beast or god was the product of careful calculation. The Roman Empire's greatest theater was not only an arena of violence. It was a stage of astonishing mechanical skill.

The Legacy of Greek Ingenuity

Rome inherited its fascination with spectacle from Greece, whose theaters used early devices to conjure the divine. The crane that lowered gods onto the stage and the rotating triangular prisms that changed backdrops were established innovations by the Hellenistic era. The Romans expanded these ideas. Their ambition for scale and drama pushed mechanical design to new heights.

By the first century BCE, Roman engineers were industrializing Greek methods. Stage machinery became part of statecraft, blending art, engineering, and propaganda. Emperors understood that control of wonder shaped loyalty. Greek theater sought to move the soul. Roman theater sought to awe it into obedience.

Mechanisms Beneath the Arena

The Colosseum was a masterpiece of engineering long before its games began. Completed under Emperor Titus in 80 CE, it held more than fifty thousand spectators. Beneath the arena lay the hypogeum, a two level system of corridors, lifts, and capstans that powered the performances above.

Dozens of elevators, some able to raise several tons, were driven by teams of workers turning massive wooden winches. Counterweights, pulleys, and guide rails kept the lifts steady as platforms shot upward through trapdoors that opened at the precise moment of each reveal. These mechanisms could be reconfigured between events, turning the arena from a hunting ground to a mythological scene in hours.

Some effects may have been hydraulic. Water pressure from aqueduct fed cisterns could have moved small floats or controlled sliding panels. Valves, adapted from Roman plumbing, may have powered fountains or quick floods that produced temporary lakes for mock naval battles.

Water, Fire, and Illusion

Romans were determined to control nature. They filled the Colosseum's lower chambers with channels and gates that allowed the arena floor to be flooded for sea battles. Ships sailed, rams collided, and floating fortresses clashed on an artificial sea. Hours later, the same arena stood dry again for gladiatorial combat.

Ancient writers describe these marvels with amazement. They also mention devices that created artificial rain and lightning, likely using pressurized air and torches behind thin screens. The blending of natural and mechanical spectacle made Rome's power appear limitless. Rome could make the earth open, the sea rise, and the heavens speak.

Fountains in imperial gardens echoed this same logic. At Nero's Domus Aurea and Hadrian's Villa, hidden waterwheels drove musical pipes and rotating statues. The principles behind these displays also powered the stage. Rome's engineering was a unified language of beauty and control.

The Theater as Imperial Instrument

Beyond the Colosseum, permanent theaters throughout the empire used machinery to raise performance to political art. Marble theaters at Pompeii, Orange, and Ephesus concealed scenery systems behind their grand façades filled with niches, doors, and balconies.

Curtains rose through slots in the stage floor, drawn by ropes and counterweights, a reverse of the modern descent curtain. Panels rotated to reveal new scenes. Rolling wagons presented tableaux of gods or fallen heroes. These mechanical transformations supported a larger purpose. When emperors sponsored plays or games, they used the stage to display imperial power. Theater became visible proof that Rome could reorder the world.

Masters of Material and Motion

Roman engineers were unmatched in practical skill. Their knowledge of wood, bronze, iron, and concrete allowed them to build machines that were strong and flexible. Pulleys and capstans used in theaters came from the same technology that hoisted stones in aqueducts and temples. Engines of war shared principles with stage lifts and trapdoors. To Romans, there was no divide between military power and theatrical wonder. Both expressed mechanical authority.

Vitruvius, in his treatise on architecture, described not only theaters but also water organs, hydraulic pumps, and lifting machines. His writing shows the

Roman belief that engineering was a branch of philosophy. It demonstrated reason's power to shape matter. This mindset made the Colosseum's feats possible.

The Machinery of Awe

A Roman festival performance celebrating an emperor's triumph might begin with trumpets as the curtain rose. From a hidden lift, a mountain appeared with mechanical trees. At its peak, an actor portraying Jupiter descended on a crane surrounded by mist. Fountains erupted as a Pegasus figure glided from one side of the stage to the other, wings moved by concealed cords. At the climax, thunder sounded, and the mountain split apart to reveal a glowing vision of Olympus.

Teams of stagehands worked in rhythm to create these illusions. They followed cues from water clocks and musicians, timing every action precisely. To Roman audiences, this coordination was more than entertainment. It demonstrated that Rome operated with the same divine precision.

Public Wonder and Private Power

Spectacle was central to Rome's political strategy. The phrase "bread and circuses" captures the idea that control of the population required both sustenance and wonder. Mechanical skill supplied the latter.

When Emperor Domitian hosted games with simultaneous gladiatorial fights, sea battles, and mythic dramas, his engineers made the impossible real. The emperor appeared as master of a world that obeyed his command as predictably as the gears beneath the arena. Every trapdoor that opened on cue reinforced the message that the empire was an ordered machine.

The Legacy of Roman Mechanics

Much Roman stage machinery vanished over time, yet its influence remained. Medieval mystery plays inherited hidden lifts and flying gods. Renaissance architects revived ancient methods to create revolving stages and self moving curtains. The great Baroque theaters, with rising platforms and rolling clouds, directly descended from Roman ingenuity.

Modern stagecraft and even theme park design preserve this legacy. Any time a stage opens seamlessly or scenery shifts with invisible precision, it echoes Roman pulleys turning beneath the Colosseum.

The Empire in Motion

The Roman passion for machinery was more than spectacle. It was ideology rendered in wood, bronze, and water. Each mechanical transformation on stage mirrored Rome's transformation of the world. The empire conquered not only nations but also nature, bending rivers into aqueducts, raising hills into amphitheaters, and turning the elements into instruments of power.

As the last performance ended and crowds left the Colosseum, the arena floor was already being reset.

Workers lowered platforms, adjusted counterweights, and closed trapdoors, preparing for the next day. Beneath the marble and sand, the gears kept turning, a symbol of the empire they served, tireless, ordered, and spectacular.

In the echo of creaking ropes and dripping water, one could almost hear the empire thinking in motion, a civilization of levers and laws where power was measured not only in armies and monuments but in the movement of a stage that made the impossible seem real.

Chapter 13
The Gears of Faith – Automata in Byzantine Churches

In the cool shadow of a marble nave, the scent of incense drifted upward in slow spirals toward a ceiling painted with golden mosaics. The congregation knelt in silence as the choir's chant echoed through the vast dome. When the final note faded, something extraordinary happened. The great icon of Christ above the altar began to turn, slowly and reverently, until its painted eyes seemed to meet those of the faithful below. A murmur rose from the worshippers. To them, it was a miracle, the living gaze of the divine made visible. Hidden behind the walls, a priest or deacon worked a system of cords and counterweights designed to move the sacred image in perfect rhythm with the liturgy.

For the Byzantines, faith and craft were inseparable. Their churches were not silent monuments but dynamic environments filled with sound, scent, light, and motion. The same civilization that preserved Greek philosophy and Roman engineering fused these traditions into a new form of devotion. Gears, levers, and hydraulics became instruments of revelation. Mechanical wonder became a language of worship.

The Legacy of Alexandria and the Language of Wonder

Centuries after the fall of Rome, the intellectual fire of antiquity continued to burn in the eastern Mediterranean. In the workshops of Constantinople, engineers inherited the writings of Hero of Alexandria, whose texts described self opening temple doors and fire powered altars. These treatises, copied in Byzantine scriptoria, became manuals for creating devices that blended spectacle with faith.

The Byzantine Empire used such mechanical arts to express mystery rather than curiosity. Hero's inventions demonstrated cleverness. Byzantine automata expressed awe. The Church understood that wonder could be a form of teaching, a bridge between human senses and eternal truth.

Moving Icons and Mechanical Liturgy

Byzantine chronicles describe icons that moved, spoke, or emitted light during services. In Hagia Sophia, visitors reported lamps that ignited without touch and censers that swung as if moved by angels. The Patriarchal throne, according to the eleventh century writer Michael Psellos, could rise or descend by hidden mechanisms, lifting the presiding bishop in a way that seemed supernatural.

Other accounts mention icons that turned toward the congregation during feast days or processions. Ropes ran through pulleys hidden behind panels, and lead counterweights rested inside columns. These mechanisms moved sacred images slowly and with ceremony, so the motion would seem organic and reverent.

In smaller chapels and monasteries, similar systems animated candle bearing angels or caused curtains to part at the moment of consecration. Worshippers did not see cords or hinges. They saw mystery. The building itself became an instrument of devotion.

The Emperor's Moving Throne

The fusion of faith and engineering reached its height in the imperial ceremonies of Constantinople. The emperor, regarded as God's representative on Earth, presided over rituals that blurred the line between earthly and heavenly power. The imperial throne room of the Chrysotriklinos contained automata that reinforced this message.

Visitors such as Liutprand of Cremona reported golden lions that roared and birds that sang when the emperor entered the room. The throne itself rose into the air on a hidden lift, startling envoys into submission. These devices, driven by compressed air, weighted levers, or hydraulic systems adapted from Alexandrian designs, turned ceremony into revelation.

Though political in purpose, their symbolism was religious. The emperor's rising throne suggested divine ascent, and the mechanical chorus of lions and birds echoed the heavenly hosts described in Scripture. Byzantine technology became theology in motion.

The Craft of the Invisible

Behind these marvels worked the mechanopoioi, the mechanical artisans. Their names rarely appear in history, yet they shaped the sensory world of Byzantine worship. Some were monks trained in metalwork and hydraulics. Others came from guilds that specialized in automata for processions or festivals.

They studied ancient knowledge from Hero, Ctesibius, and Archimedes, and adapted it for sacred purpose. Their work was a theology of motion. A gear turning behind an icon was not merely a machine. It was a symbol of divine power acting through matter.

Unlike many Western artisans, the Byzantine craftsman aimed for invisibility. His highest achievement was to make his work vanish, leaving only wonder behind. If worshippers believed an icon had turned by divine will, he had succeeded.

Sound, Light, and the Architecture of Wonder

Byzantine worship engaged all the senses. Mechanics enhanced each one. Bronze and silver censers suspended on long chains swung through the air by means of oscillating counterweights. Oil lamps lowered for refilling and then rose again, balanced on silent pulleys.

Light itself became sacred motion. The dome of Hagia Sophia, pierced by forty windows, appeared to float on gold at dawn. When lamps and candles moved in coordinated sequences, the building seemed alive.

Acoustics added to the effect. Some churches embedded bronze jars in their walls to amplify sound, following an ancient technique borrowed from Greek theaters. Chant echoed with depth. Mosaics shimmered. Automata moved in measured rhythm. Everything worked together to evoke the harmony of the heavens.

Faith and Physics

To modern eyes, these devices may seem like illusions. To the Byzantines, they revealed truth. In their worldview, natural law and divine law were reflections of each other. If God had created a universe governed by order, then human craftsmanship that reproduced that order became a form of devotion.

Theologians spoke of theurgy, a practice in which human action made divine presence perceptible. Mechanical motion became a kind of prayer. The hidden gear behind an icon served the same purpose as the hymn. Both revealed the same mystery.

The Byzantine cosmos was a living mechanism maintained by God. Human engineers, in their small way, imitated His wisdom.

The Transmission of the Sacred Machine

When Constantinople fell to the Fourth Crusade in 1204, Western visitors carried stories of its wonders back to Europe. Chronicles from Venice and Sicily describe "singing birds" and "moving thrones" taken from the imperial palace. These marvels may have influenced later medieval automata built for Western cathedrals. Clockwork angels rang bells. Mechanical Magi bowed before the Virgin. Crucifixes moved their eyes during mass.

Through these channels, Byzantine mechanical devotion shaped the Gothic imagination. Wonder once again became a path to faith.

The Liturgy of Motion

In the Byzantine church, movement revealed presence. Every turn of an icon and every swing of a censer reminded the faithful that the divine was near. The line between miracle and mechanism disappeared. The two became one language.

As the final hymn faded, the icon above the altar completed its slow arc and faced forward again. The congregation crossed themselves, believing they had witnessed a sign. Behind the walls, a counterweight descended, a pulley turned, and human craft gave shape to faith.

In that quiet partnership between devotion and design, the Byzantines achieved something enduring. They created a vision of worship where art, science, and belief moved together in harmony. Each motion was a prayer. Each mechanism offered a glimpse of the divine order at work in the world.

Chapter 14
The Artisans of Illusion – Courtly Engineers of the Islamic Golden Age

The air in the sultan's garden was thick with the scent of jasmine and oil lamps. Beneath a canopy of silk, courtiers reclined on cushions, sipping sherbet from crystal goblets as musicians played soft melodies. Suddenly, a golden peacock at the center of the fountain began to move. Its head turned with slow grace. A whisper of gears stirred within the pedestal, and with an elegant shimmer, the bird's tail unfolded into a fan of jeweled feathers. Water cascaded from its beak, glinting like liquid silver in the moonlight. The guests gasped. Their host smiled. The device had done its work, not only as entertainment but as a demonstration of dominion over knowledge itself.

In the courts of the Islamic Golden Age, illusion became a language of power. From Baghdad to Damascus, engineers known as mechanical philosophers built marvels that blurred the line between art and science, faith and reason. Their creations, including automated fountains, self playing instruments, and lifelike birds and animals, reflected a world that saw curiosity as devotion. Each device was both spectacle and statement, revealing how intellect and craft could mirror the harmony of creation.

The Splendor of the Caliph's Court

The ninth century marked the height of Abbasid power in Baghdad, a city known as Madinat al Salam, the City of Peace. Its caliphs gathered the greatest minds of their age at the Bayt al Hikma, the House of Wisdom, where Greek manuscripts were translated into Arabic and refined through experiment. Mathematics, astronomy, and mechanics intertwined to shape a new vision of beauty grounded in order.

In this intellectual climate, the Banū Mūsā brothers, Muhammad, Ahmad, and al Hasan, flourished. Mathematicians and inventors, they combined geometry, hydraulics, and pneumatics to create what they called ingenious devices. Their book, Kitāb al Ḥiyal, described a hundred mechanical designs: fountains that changed shape, self trimming lamps, and water organs that played without human touch. These inventions were not toys. They were proofs of understanding, demonstrations of mastery through proportion and flow.

The Banū Mūsā and the Language of Flow

One of their most admired creations was a self operating fountain capable of altering its pattern automatically. By directing water through cleverly designed valves and air chambers, the brothers produced jets that rose, fell, and curved in rhythm. To the caliph and his guests, it seemed as if the water obeyed music.

Another marvel, a wine dispenser, mixed liquids in precise ratios before serving them through hidden tubes. Such devices delighted patrons while also revealing mathematical insight. Every pipe, lever, and siphon embodied geometric logic, a living expression of algebra in motion.

The Banū Mūsā understood that wonder was not opposed to reason. As they wrote, "Those who see it will imagine it is by magic, but it is by geometry." Their machines made science visible, turning calculation into spectacle.

The Rise of Al Jazari

Three centuries later, in the thirteenth century, Ismail Al Jazari carried their legacy forward. Working in the court of the Artuqid rulers of Diyarbakır, he inherited both the theory and the spirit of the Banū Mūsā. His Book of Knowledge of Ingenious Mechanical Devices catalogued fifty inventions that transformed engineering from craft to discipline.

While the Banū Mūsā emphasized fluid dynamics, Al Jazari married science to artistry. His peacock fountain, perhaps the ancestor of the scene that opened this chapter, was both functional and beautiful. Water flowed from a bronze bird into a basin, activating floating cups that rose, poured, and reset themselves in a continuous loop. The result was an illusion of perpetual life, a mechanical garden of paradise.

He also designed automatic servants that poured water for ablutions and reset themselves after each use through levers and counterweights. His most famous works, including the elephant clock and musical automata, combined intricate gearing with refined craftsmanship. Gold, silver, enamel, and carved wood adorned his mechanisms. They were machines elevated into art.

The Engineer as Philosopher

In Islamic society, the engineer was not merely a technician. He was a philosopher of nature. Mechanics reflected divine order, a discipline that revealed how God's universe moved through harmony and precision. The Qur'anic idea that "everything is measured" inspired scholars to find mathematical balance in all things, from architecture to hydraulics.

For Al Jazari and his predecessors, mechanical art expressed tawhid, the unity of creation. Their devices were not secular amusements. They were sermons in brass and water.

Wonders for the Court

The courts of the Abbasid and later Seljuk rulers were fertile ground for such marvels. Caliphs surrounded themselves with devices that embodied refinement and divine favor. A throne that rose through hydraulic force symbolized elevation through God's will. A mechanical tree filled with singing birds represented paradise, the ultimate reward for the faithful.

Contemporary accounts describe automated gardens built for princes in Damascus, where mechanical birds flapped and artificial fish swam in marble basins. Waterwheels turned music boxes hidden within the walls, creating a symphony of sound and motion. Visitors wandered among these wonders in astonishment, praising both the engineer's skill and the ruler's wisdom.

Technology became theater. The machines were carefully timed to surprise their audience. But rather than concealing ignorance, they revealed understanding. To witness such a performance was to glimpse the intellect of a civilization at its height.

Craft and Collaboration

These automata required collaboration between artisans of many trades. Metalworkers, carpenters, calligraphers, and mathematicians worked together. Gears were cut from brass by hand. Leather seals prevented leaks. Inscriptions were engraved in elegant Kufic script. Each creation was both science and scripture, precise in action and poetic in form.

Workshop records suggest that engineers trained apprentices to maintain and adjust the machines. These students learned to think mechanically, developing skills that later influenced European engineering. Through Andalusia and Sicily, Arabic mechanical principles reached Latin scholars, shaping the automata of Renaissance Italy.

The Culture of Wonder

In the Islamic Golden Age, wonder was considered a form of wisdom. The philosopher Al Kindi wrote that marvels, when understood, bring one closer to truth. To provoke astonishment was to invite contemplation. A caliph who owned an automaton did not merely display wealth. He demonstrated participation in cosmic intellect.

This view distinguished Islamic mechanical art from later European work. In Europe, automata often symbolized man's challenge to nature. In Islam, they expressed harmony with it. A device that imitated life was not rebellion but tribute.

Legacy and Rediscovery

After the Mongol invasions and the decline of great courts, many manuscripts and devices were lost. Yet their memory endured. In the fourteenth century, translations of Arabic mechanical texts reached Venice and Paris, where they inspired Giovanni de Dondi, Leonardo da Vinci, and other Renaissance engineers. Al Jazari's influence appears in the escapements and automata of early clocks and in fountains that mimic his flowing choreography.

Modern historians recognize these courtly engineers as early ancestors of automation. Their devices joined elegance with insight, entertainment with education. They proved that precision and poetry could coexist, and that machinery could express intellect and emotion.

The Eternal Performance

As the golden peacock's feathers folded and the fountain resumed its quiet flow, the sultan's guests applauded softly. None saw the gears turning below or the careful balance of weights and siphons that gave life to metal. But that was the point.

In the courts of the Islamic Golden Age, the greatest illusion was harmony. The peacock, the fountains, and the singing birds reminded all who watched that beneath the visible world lay a deeper order, revealed through geometry, artistry, and faith.

As the last droplets of water caught the lantern light, the peacock grew still. Somewhere within its pedestal, a small valve closed, ready for the next cue. The illusion ended, yet the idea remained: beauty, like knowledge, was a mechanism crafted to move the soul as surely as a gear turns the world.

Chapter 15
The Pulse of Time — The Legacy of Ancient Clocks

In the predawn stillness, a single droplet fell from a clay vessel and disappeared into a shallow basin. Another followed, then another, each tiny splash echoing through the darkness. The rhythm was steady and unbroken, a whisper of order in the stillness before sunrise. For the priests of ancient Egypt, those drops measured not only the passing of hours but the harmony of creation. Long before sundials marked the sky or sand flowed through glass, humanity learned to mark existence in falling water.

From the banks of the Nile to the observatories of China and the mosques of Baghdad, the story of the water clock is the story of civilization's heartbeat. It is the record of humankind's oldest mechanical pursuit, the desire to transform motion into meaning.

The Egyptian Invention

The earliest known water clocks appeared in Egypt more than three thousand years ago. At Thebes, in the temple of Amun at Karnak, archaeologists have uncovered limestone vessels marked with hieroglyphic scales that measured hours through the night. Water flowed from a small hole at the bottom, and as the level fell, priests read the hours from the markings inside the bowl.

The Egyptians divided the night into twelve equal parts, adjusting their markings for the changing length of seasonal hours. The flow of water required careful calibration. Temperature and viscosity could distort the rhythm, and so precision became a sacred task.

These clocks served ritual rather than daily needs. Time belonged to the gods, not to the marketplace. The drip of water in a darkened temple chamber echoed the eternal flow of the Nile and the renewal it brought. The priests who tended the clepsydra were maintaining not just time but cosmic balance.

The Babylonian Rhythm

To the Babylonians, time was celestial. Their astronomers mapped the stars with exacting care, searching for patterns that connected heaven and Earth. Water clocks became astronomical tools, measuring how long stars remained visible and how quickly the moon crossed the sky.

Cuneiform tablets describe bowl shaped clepsydrae marked inside to allow astronomers to measure intervals crucial for predicting planetary motion. These timekeepers tied human perception to the heavens. They transformed observation into calculation and calculation into prophecy.

The Greek Transformation

In Greece, the water clock became a machine. The word clepsydra means "water thief," and Greek inventors did not disappoint. By the fifth century BCE, creators like Ctesibius of Alexandria were building devices that used pressure and siphons to control the flow.

Ctesibius's regulated clepsydra introduced feedback mechanisms. Floating indicators connected to rods maintained a steady flow despite fluctuating water levels. This was more than innovation. It was the earliest self correcting system, the ancestor of feedback control and automated regulation.

In Athens, water clocks also had civic duties. Courts used them to time speeches. As an orator presented his case, water dripped steadily from a bronze urn. When the last drop fell, his time expired. Justice and rhetoric moved to the same invisible cadence.

The Hellenistic Expansion

Hero of Alexandria later transformed the clepsydra into spectacle. He connected water flow to moving statues, rotating dials, and automatic chimes, turning practical devices into small theaters of motion. These were not clocks in the modern sense but choreographed demonstrations that joined measurement with wonder.

For the Hellenistic world, beauty and function were inseparable. In their hands, a clock became both instrument and inspiration.

The Chinese Innovators

Far to the east, timekeeping in China evolved on monumental scale. As early as the Han dynasty, engineers built dripping water clocks that tracked celestial events with notable accuracy. By the eighth century, Yi Xing, a Buddhist monk and astronomer, created a clock that used water and escapement mechanisms, anticipating later European designs.

The pinnacle came in the eleventh century with Su Song's clock tower in Kaifeng. Nearly thirty feet tall, it used a chain drive powered by water to rotate an armillary sphere. Figures emerged from doors to strike bells marking the hours.

Su Song's tower was more than a machine. It was a declaration of cosmic order. The emperor, as keeper of harmony between heaven and Earth, relied on such mechanisms to demonstrate his role. The tower united observation, mathematics, and artistry into one of the world's most advanced medieval machines.

The Islamic Golden Age

When Greek and Persian learning reached Baghdad in the eighth and ninth centuries, water clocks became symbols of enlightenment. The Banū Mūsā brothers studied ancient designs and created intricate fountains and timing devices that combined water, air, and balance weights.

Their contemporary Al Jazari perfected the art. His clocks did more than measure time. They narrated it. His elephant clock, for example, used a water timed float hidden inside the elephant's body. Every half hour a figure dropped a ball, triggering chimes and automata representing different cultures: an Indian elephant, a Persian carpet, an Egyptian phoenix, a Chinese dragon.

The symbolism was deliberate. Al Jazari's clock was a mechanical map of the known world, united by the rhythm of water. Each drop expressed a shared pulse of human ingenuity.

Time as Divine Geometry

Across civilizations, timekeeping was sacred. To measure time was to approach the structure of creation. Egyptians aligned temples to the stars. Babylonians mapped celestial cycles. Islamic scholars calculated prayer times through geometric astronomy.

Cultures expressed this respect in the beauty of their clocks. Greek clepsydrae stood in marble courtyards carved with waves. Chinese clocks gleamed with lacquer and bronze dragons. Arabic fountain clocks paired geometric tiles with poetry. Each expressed a truth older than science: to understand time was to glimpse eternity.

From Flow to Gear

Eventually, the pulse of time changed. After Arabic engineering spread through Europe, artisans replaced water with mechanical escapements and gears. By the fourteenth century, tower clocks rose in city squares. The sound of dripping water gave way to iron bells.

Yet even these clocks carried ancient logic. The idea that motion could be regulated—first by water, then by weight, then by springs—was a direct legacy of hydraulic intelligence spanning continents.

Every modern tick echoes a drop falling from a clepsydra, a whisper from a temple hundreds of generations old.

The Human Heartbeat

The evolution of the water clock reflects a universal need to find rhythm in the world. Water was chosen because it moved like life itself, fluid and continuous. To measure it was to impose reason upon motion, to give form to the intangible.

From priests counting drips in moonlit stone chambers to emperors watching celestial spheres, the pursuit of time has always been an act of wonder. Each culture shaped its clocks according to its vision of order, yet all shared the same impulse: to bring the cosmos within reach.

Even now, with digital displays that measure to the millisecond, we remain heirs to that ancient desire. Time flows, whether through water, gears, or light, and we still try to catch it, to see in its passage our own reflection.

In imagination, the clay vessel still drips. Each fall of water marks a heartbeat of understanding. It reminds us that technology began not with conquest but curiosity, with the quiet effort to listen to the world's rhythm and give it measure. The pulse of time began in a temple basin, and it has never stopped.

Chapter 16
The Scrolls That Spoke – Lost Manuscripts of Mechanism

In a cold monastery scriptorium, the faint light of an oil lamp flickered across stone walls. A monk bent over a rolled parchment unlike any he had ever seen. Its margins were filled with graceful Arabic script and drawings precise and strange, showing gears interlocked like teeth, bellows feeding pipes, and a vessel balanced on shifting weights. He traced a finger over one image, a bird that sang through hidden pressure, and whispered in awe, "The scroll speaks."

The document before him was a relic of a forgotten world, one in which engineers had built living machines, fountains that danced, and clocks that told the hours through motion and music. Although he could not read all its words, the diagrams captured his imagination. The scroll, a survivor of centuries and conquests, carried within it the mechanical dreams of civilizations long past.

This is the story of how those dreams traveled, how the knowledge of automata and ancient mechanics, nearly lost to war and neglect, was preserved by scribes, translated by scholars, and reborn in new languages across continents.

The Survival of Greek Genius

When the Western Roman Empire crumbled, much of classical science vanished from Europe's libraries. Scrolls once copied in Athens and Alexandria were burned, buried, or forgotten. Yet Greek learning lived on in the eastern Mediterranean and in the great cities of the Islamic world. Manuscripts by Ctesibius, Hero of Alexandria, and Archimedes continued to circulate.

Hero's *Pneumatica*, which described self opening temple doors, coin operated machines, and theatrical automata, was especially influential. Its precise language of pipes, valves, and siphons became a foundation for mechanical experimentation for more than a thousand years. Even as Greek originals faded, Arabic scholars translated and expanded these ideas, preserving not only the inventions but the spirit behind them.

The Translation Movement

In the ninth century, the House of Wisdom in Baghdad became a crossroads of knowledge. Scholars sought Greek, Persian, and Indian works to translate into Arabic. The *Pneumatica* of Hero, the mechanical writings attributed to Aristotle, and treatises by Archimedes all passed through this intellectual gateway.

Translators such as Hunayn ibn Ishaq and Thabit ibn Qurra rendered these texts into Arabic and added commentary that clarified and sometimes corrected ancient ideas. Their work did more than preserve the past. It transformed it. Concepts dormant in Greek manuscripts emerged with new precision in the mathematical language of Arabic science.

Engineers like the Banū Mūsā brothers and Al Jazari did not merely copy the ancients. They continued the tradition. Their books of ingenious devices extended Hero's principles into new forms powered by water, air, and weights. These works later astonished European monks and scholars who saw them centuries after they were written.

The Journey West

Knowledge travels with merchants, pilgrims, and conquerors. In the twelfth century, as Europe rekindled its intellectual curiosity, the translation movement shifted to Toledo, Sicily, and Salerno. These were regions where Islamic, Christian, and Jewish cultures met.

In Toledo, archbishop Raymond of Sauvetat sponsored teams of translators who rendered Arabic scientific works into Latin. Their workshop became a bridge across centuries. Treatises on optics, astronomy, and mechanics passed from Arabic into the language of medieval Europe. Among these were portions of Hero's *Pneumatica* and parts of Al Jazari's *Book of Knowledge of Ingenious Mechanical Devices*.

In Sicily, under Frederick II, Arabic manuscripts circulated freely. The emperor's court welcomed engineers, philosophers, and translators. Latin accounts of Frederick's automata, including mechanical fountains and singing birds, resemble designs from Al Jazari's book.

The Voices of the Scrolls

These manuscripts carried more than diagrams. They carried philosophy. They revealed that nature could be imitated through proportion and that mechanism could express order rather than defy it. Each gear and lever represented an invisible harmony made visible through craftsmanship.

Monks in monasteries such as Cluny or Monte Cassino saw mystery in these diagrams, even when they did not understand every principle. They recognized that motion could be governed by reason. The scrolls spoke in the language of geometry and spiritual reflection.

Gradually, these ideas inspired new European experiments. Monastic inventors built water clocks and astrolabes. Cathedral craftsmen created angels that moved and bells that rang without human touch. The automaton returned, now a Christian marvel symbolizing divine precision.

The Hidden Renaissance Before the Renaissance

By the late Middle Ages, the influence of these manuscripts appeared across Europe's learning centers. Scholars in Oxford, Paris, and Padua studied mechanics as a science. Roger Bacon praised the Arab translators for restoring ancient wisdom. He imagined flying machines and self moving carts—dreams shaped by descriptions of Hero's devices.

In Catalonia and Italy, engineers built intricate clockworks influenced by Arabic and Byzantine mechanisms. Giovanni de Dondi's fourteenth century astronomical clock owed much to these translated designs. Every wheel in its gear train echoed a lineage stretching back to Alexandria.

Even Leonardo da Vinci later sketched hydraulic lifts and musical automata that mirrored Al Jazari's work. He may never have read the Arabic texts directly, but their descendants shaped the world around him.

The Fragility of Memory

Many of these manuscripts have been lost again, scattered through war, decay, or quiet neglect. Yet fragments still surface. A torn page of Hero's *Pneumatica* in an Istanbul library. A Latin translation of Al Jazari found in Florence. A marginal note from a twelfth century scribe wondering at a "vessel that moves itself."

These fragile survivors remind us that civilization advances through rediscovery. Knowledge passes like flame from one wick to another, sometimes dimming, sometimes blazing.

The Mechanical Imagination Reborn

By the Renaissance, these rediscovered manuscripts transformed European thought. Engineers treated machines as instruments of understanding. Clockmakers studied ancient water pressure and escapement designs. Architects borrowed geometry from ancient fountains. Artists filled sketchbooks with gears and pulleys inspired by the old diagrams.

In this rebirth, art and science merged again. Just as Hero's machines had guided the hands of temple builders, Al Jazari's diagrams inspired workshops across Europe. The mechanical imagination, once confined to Greek theaters and Islamic courts, became a universal language.

The Eternal Scroll

As the monk in the scriptorium gently re rolled the parchment, the candle beside him dimmed. He did not know that centuries later others would unroll similar scrolls and marvel at the same lines and circles. The diagrams would travel again, from ink to blueprint to printed page, each age translating them anew.

The scrolls that once spoke in Greek, then in Arabic, now whisper in every machine that hums, every clock that marks the hours, every unseen gear that turns. They remind us that knowledge is not bound by borders or time. It flows like water, carrying the ingenuity of the ancients through the rivers of history.

In every rediscovery lies a resurrection, and in every turning wheel, an echo of those who first dreamed of motion made eternal.

Chapter 17 The Return of the Machines — Renaissance Rediscovery

The light of a single candle trembled over a table strewn with sketches. Ink stained fingers moved swiftly, tracing circles within circles, gears nested in frames, and levers linked by cords of thought. Leonardo da Vinci paused, glancing from the parchment before him to a group of ancient looking texts, Latin translations of works that had traveled through Byzantium and Baghdad from their origins in Alexandria. On one sheet, a bronze knight seemed ready to stand. On another, a system of pulleys and counterweights mimicked the rhythm of a beating heart. Around him, the room seemed to hum with potential, the Renaissance mind awakening to the mechanical past.

The Age of Reawakening

By the fifteenth century, Europe had entered an age of rediscovery. Libraries, monasteries, scholarly archives, and eastern trade routes carried forgotten knowledge back into circulation. Greek and Arabic manuscripts, many of them translations of Hero of Alexandria, Ctesibius, and Al Jazari, were studied by engineers and artists who recognized science rather than superstition in their pages.

In Florence, Venice, and Milan, inventors and patrons believed that the mechanical arts could express divine order through design. The Renaissance was not only a revival of art and learning. It was a revival of motion, the belief that beauty could move through the precision of gears and gravity.

The Inheritance of Hero and Al Jazari

When manuscripts of Hero's *Pneumatica* and *Automata* resurfaced in Western Europe, they captivated the imaginations of engineers. These works described self opening temple doors, air powered fountains, and mechanical theaters capable of performing entire plays through timed sequences. Though centuries old, they felt strikingly modern to Renaissance readers.

Fragments of Arabic mechanical treatises, especially those by Al Jazari, also circulated through Italy by way of Sicily and Andalusia. Drawings of water clocks, automata, and ingenious pumps appeared in compendiums studied by Giovanni Fontana and Taccola, predecessors of Leonardo. These men viewed mechanical arts as a branch of natural philosophy. Where earlier thinkers sought divine causes, they sought mechanical ones, finding divinity in order itself.

Leonardo and the Mechanical Renaissance

Among those who inherited this legacy, Leonardo da Vinci stands foremost. His notebooks reveal a mind that combined observation, geometry, and imagination. To Leonardo, the machine was a living organism, each component like bone, muscle, or sinew, each force reflecting the same principles that governed the human body.

His mechanical knight, designed around 1495, embodies this vision. It could stand, sit, and move its arms through internal pulleys and gears. Leonardo's sketches show a clear understanding of linkages, weights, and rotary motion, principles that had appeared in Hero's diagrams fifteen centuries earlier. Whether he encountered Hero's works directly or through intermediaries, Leonardo reinterpreted them with new ingenuity.

Leonardo also explored hydraulic mechanisms that echoed Al Jazari's fountains and self regulating water systems. He drew designs for self filling basins, automatic spigots, and floating indicators that measured water flow. His ambition reached beyond imitation. He sought to understand how motion could be harnessed, balanced, and beautified.

Giovanni Fontana and the Revival of Automata

Before Leonardo's rise, the Venetian engineer Giovanni Fontana had already begun rebuilding classical automata. In his *Bellicorum Instrumentorum Liber*, written around 1420, Fontana illustrated walking skeletons powered by hidden wheels, flying dragons suspended by cords, and fountains animated by concealed pipes.

Fontana's notes show that he had access to Latin translations of Hero and Arabic works on hydraulics. He was one of the first to experiment with magnetism, air compression, and optical illusion, blending ancient mechanism with the new interest in physics. His automata symbolized mastery over natural law rather than mere spectacle.

Taccola and the Engineering Humanists

Another Renaissance engineer, Mariano di Jacopo, known as Taccola, expanded this revival. Working in Siena, he produced treatises filled with drawings of mechanical lifts, rotating stages, and hydraulic systems inspired by classical designs. His influence reached Francesco di Giorgio Martini, who later mentored Leonardo.

These men formed a chain of transmission linking rediscovered manuscripts with Renaissance creativity. Their drawings carried Hero's and Al Jazari's ideas into architecture, city planning, and performance. The ancient engineers had built for temples and courts.

The Renaissance engineers built for cathedrals, theaters, and cities.

Motion as Art

To Renaissance thinkers, motion became an aesthetic principle. Painters studied perspective. Sculptors sought the illusion of movement. Engineers animated their inventions. The automaton, once a temple marvel or royal toy, became a metaphor for life, intellect, and harmony.

Leon Battista Alberti wrote that beauty arises from proportion and purpose. His words echoed the belief that governed Hero and Al Jazari, that the laws of water flow or gear ratios reflected a larger cosmic order. To design a perfect mechanism was to imitate creation itself.

The Spectacle of Ingenuity

Mechanical displays became popular across Europe. In Florence and Nuremberg, clockwork figures of saints and angels moved with precision. In Venice, engineers staged mechanical theaters that reenacted mythological scenes with hidden cranks and pulleys.

These spectacles resembled performances in the theaters of Alexandria and the temples of Byzantium, yet their meaning had changed. The Renaissance automaton celebrated human genius rather than divine mystery. Life like motion, once a sign of divine power, now testified to human intellect.

The Birth of Scientific Mechanics

As mechanical arts gained respect, they also gained precision. Universities studied motion mathematically, merging invention with theory. The rediscovery of Archimedes' work on equilibrium and Hero's principles of pneumatics formed the foundation of early mechanics.

Renaissance engineers developed models for levers, pulleys, and gears that anticipated Newtonian science. Leonardo's diagrams of motion and resistance read like experiments, blending artistry with quantitative thought. His pursuit of perpetual motion machines, though impossible, reflected the same curiosity that inspired Hero's water driven devices.

A Legacy Reawakened

By the sixteenth century, Europe's workshops and studios had become laboratories of invention. Clocks grew more precise, fountains more elaborate, and architecture more ambitious. The spirit of the ancient engineers had returned, transformed by Renaissance imagination.

Even as printing spread knowledge rapidly, the old handwritten diagrams remained treasured. Leonardo and his peers saw them not as relics but revelations. Ancient engineers had dreamed of self moving statues. The Renaissance made them plausible.

The Continuum of Motion

In the candlelight, Leonardo's pen traced one last line before he closed his notebook. The sketch before him, half man and half machine, seemed almost alive. It was both homage and prophecy, a bridge between the world of gears and water and the coming age of metal and steam.

The rediscovery of mechanical knowledge during the Renaissance did more than revive forgotten texts. It reignited humanity's sense of possibility. From Hero's temples to Al Jazari's courts to Leonardo's studio, the thread of imagination had never broken. It had only waited to be rewound.

As the candle dimmed and the room darkened, the knight on the page remained poised between memory and invention. The ancients had shown how to make matter move. The Renaissance, rediscovering their secrets, began to make thought move as well.

Chapter 18
The Spirit of Invention – How Ancient Automata Inspired Modern Robotics

In a quiet museum laboratory, engineers bend over a gleaming bronze sphere connected to a tangle of glass tubes and valves. A small flame flickers beneath a chamber of water, and steam begins to hiss through a narrow nozzle. The sphere starts to spin. It is Hero of Alexandria's aeolipile, the world's first steam engine, rebuilt two thousand years after its invention. Cameras click. One researcher murmurs, "He almost had it." Around them, replicas of other ancient devices hum softly: Al-Jazari's water clocks, a mechanical bird that sings through air pressure, and a bronze hand that opens when warmed by a candle. The room feels reverent, as if the ancients themselves were present, watching the future rediscover their dreams.

The modern age of robotics and artificial intelligence did not begin in isolation. Its roots reach back to early experiments in motion, when priests, philosophers, and engineers first wondered whether the spark of life could be imitated through mechanism. Today's algorithms and circuits are the distant descendants of bronze gears and dripping water. They are testaments to the enduring human desire to give thought and spirit to matter.

The Echo of the Ancients

Every era imagines the future in the image of its past. For the Greeks and Egyptians, movement signified divinity. For the modern world, it often signifies intelligence. The parallels are striking. Hero's temple doors that opened by fire pressure foreshadow the automated sliding doors of airports. Al-Jazari's self-operating fountains resemble modern hydraulic feedback systems. Su Song's clock tower in eleventh century China, powered by a chain drive and escapement, carries the mechanical DNA of robotics.

The idea that a machine could imitate life fascinated early thinkers long before electricity or code. Aristotle speculated that if tools could perform tasks on their own, "masters would have no need of slaves." Automation has fulfilled that vision in ways he never imagined. Yet the spirit behind it remains the same: curiosity about the boundary between nature and invention.

A Renaissance of Reconstruction

In workshops and universities around the world, engineers now recreate the automata of antiquity to understand how they worked and why they mattered. The Hero Project at the University of Thessaloniki has rebuilt nearly all of Hero's described devices using historically accurate materials, including bronze siphons, leather valves, and glass pumps. When these machines perform, doors swing open, birds chirp, and statues move. Audiences experience wonder similar to what ancient worshippers must have felt.

Robotics labs also study Al-Jazari's mechanical principles. His feedback systems, where one motion regulated another, anticipate the logic of closed loop control. This logic is foundational to robotics. Even the simple float valve he used in his fountains survives today in household plumbing, industrial pumps, and climate control systems. The continuum from medieval craftsmanship to modern engineering is not symbolic. It is mechanical and direct.

From Motion to Mind

Ancient automata raised questions that still challenge roboticists and philosophers of artificial intelligence. Where does motion end and thought begin? Is lifelike imitation merely mechanical, or can it reveal something about consciousness itself?

The ancients framed these questions in religious and mythic terms. The moving statues of Egypt or Byzantium symbolized divine power made visible. The myth of Talos, the bronze giant who patrolled Crete, warned against pride in creation. Today, debates about artificial intelligence echo similar concerns. When engineers teach a robot to walk, or to compose music, they reenact a very old drama: the effort to bridge the space between the living and the made.

In every age, the act of animation carries deep meaning. Whether through pneumatics or programming, the impulse to make matter move remains an expression of belief in knowledge, possibility, and human creativity.

The Invention of Autonomy

The step from ancient automaton to modern robot lies in autonomy, the ability to act without direct human input. Yet ancient engineers hinted at independence. Hero's devices responded to fire, air, and water, performing tasks once the initial condition was set. They were early programs, executing sequences based on physical logic. Al-Jazari's clocks performed routines for hours. Su Song's celestial models rotated with mechanical regularity, imitating the heavens.

Modern robotics extends this autonomy from mechanics to computation. Instead of water flow and pressure, the forces governing motion are electricity and algorithms. A self balancing robot uses feedback sensors just as a float valve maintains stable water. Machine learning reflects the ancient engineer's testing and adjustment, refining weights and pressures until harmony is achieved. The philosophy of robotics began when the first craftsman noticed that cause and effect could be designed.

Creativity and the Machine

The ancient automaton was a marvel. The modern robot often stirs anxiety. Yet both provoke the same question: What does it mean to create something that appears to think or act on its own?

When a modern artificial intelligence composes a melody or paints an image, observers argue over whether the act is creative. The same doubts surrounded Hero's devices. Did they perform miracles or clever illusions? For temple visitors, the difference mattered little. The wonder was real. Likewise, when a humanoid robot speaks or plays an instrument, the line between artifice and intention blurs. In both eras, what moves us is the illusion of agency, the sense that something beyond simple motion is taking place.

The Ethics of Invention

Ancient philosophers warned that invention without wisdom could lead to danger. The myth of Daedalus cautioned against ambition. The myth of Pygmalion questioned whether a creation could replace its creator. Today, artificial intelligence prompts similar caution. If a robot acts independently, who bears responsibility? If an algorithm causes harm, who is accountable?

The ancients framed these questions through myth. We frame them through ethics and law. Yet the struggle remains unchanged: how to balance creativity with wisdom.

Faith in Motion

Many museums display reconstructed automata beside interactive robotics exhibits, inviting visitors to compare ancient mechanics with digital intelligence. Standing before Hero's whirring devices or Al-Jazari's elephant clock reveals a common lineage of wonder. The drive to animate the inanimate is not about utility alone. It is about transcendence.

Modern engineers often describe a robot's first successful movement as miraculous. A priest in Alexandria or a scholar in Baghdad would have understood that sentiment. To create motion where none existed is to participate, however briefly, in the mystery of creation.

The Eternal Thread

From clay and bronze to silicon and code, the materials have changed, but the intention has not. Humanity continues to build machines that reflect its own image, extensions of curiosity, artistry, and longing. The robotics lab, with its blinking lights and humming servos, is the descendant of ancient temple workshops filled with bronze gears and burning lamps.

Each new invention, from automated prosthetics to learning algorithms, carries echoes of the past. The first engineers taught us how to harness nature's laws. The new ones teach us how to teach machines. Both seek to understand humanity by recreating what we are.

The Machine That Dreams

As the aeolipile spins and slows, steam condensing into silence, the researchers smile. They are witnessing not only the birth of a machine but the continuation of an idea. From Hero's fire powered devices to the neural networks of artificial intelligence, the thread of invention remains unbroken.

In the reflection of bronze and glass, past and present blur. The ancient engineers sought divinity through mechanism. Modern engineers seek humanity through design. Both chase the same dream: to bring stillness to life, to make the universe respond.

The machines have returned, no longer curiosities but companions in creation. In their motion, we can still hear ancient voices whisper through time: Build, and you will understand.

Chapter 19
The Silence of the Machines — The Disappearance of Ancient Engineering

Across the drifting sands of an Egyptian ruin, wind sighs through hollow stones once filled with pipes and valves. Bronze fragments glint faintly in the sunlight, a bent gear, a corroded lever, the curved beak of a bird that once sang through hidden air. The fountain is long dry, its voice silenced by centuries. Statues that once moved now stand motionless, their secrets buried beneath dust and time. Somewhere underfoot, the remnants of a civilization's ingenuity sleep, forgotten not because it failed, but because the world changed around it.

The silence of the machines is one of history's quiet mysteries. How could societies that built automatic doors, self filling basins, and mechanical theaters lose such knowledge for more than a thousand years? The answer lies not in a single catastrophe but in a slow unraveling, shaped by war, secrecy, shifting priorities, and the fragile nature of memory.

The Fragility of Knowledge

Ancient engineering was recorded on materials not meant to last. Papyrus, parchment, wax tablets, and oral instruction carried the wisdom of entire eras. The machines themselves, built from bronze, wood, and iron, were vulnerable to corrosion and repurposing. When kingdoms fell, devices were dismantled for scrap. Manuscripts were burned in conquest or lost in the chaos of collapse.

The Library of Alexandria, which held treatises on mathematics, mechanics, and astronomy, suffered several fires between 48 BCE and the third century CE. Each disaster erased a chapter of human ingenuity. Among the losses may have been works by Ctesibius and Philon of Byzantium, whose writings formed the early backbone of mechanical science. Their machines, described in texts no longer preserved, were dismantled by invaders or forgotten by descendants who no longer understood how they worked.

Even surviving manuscripts faced the danger of misinterpretation. A single missing measurement or damaged diagram could render an entire device useless. Without constant practice, knowledge fades into silence.

The Fall of Empires

Every empire that nurtured mechanical innovation eventually crumbled. When Rome collapsed in the fifth century, its workshops and foundries went dark. The knowledge of hydraulic systems and stage machinery faded with the empire that had funded them. In the east, the Byzantine Empire preserved fragments of the tradition, yet political instability and theological caution limited open experimentation.

The Islamic Golden Age inherited and expanded Greek and Persian science, but its centers also faced destruction. When the Mongols sacked Baghdad in 1258, they destroyed the House of Wisdom, the heart of Arabic scholarship. Chronicles claim that so many books were thrown into the Tigris River that its waters ran black with ink.

China faced similar disruptions. The great eleventh century clock tower of Su Song in Kaifeng was dismantled by invading Jurchen forces. Its parts vanished, and its advanced chain drive would not appear again in any civilization for centuries.

Knowledge, like the empires that support it, is mortal.

The Veil of Secrecy

Another force that silenced ancient machines was secrecy. Engineers served rulers, temples, and patrons who valued exclusivity. A priest using hidden hydraulics to animate a statue guarded his methods as sacred mystery. A court mechanic who built marvels for a caliph protected his craft as a state secret.

Hero of Alexandria hinted at this in his writings, noting that some temple devices were known only to initiates. Revealing their mechanisms would spoil the illusion, and with it, the power of belief. This secrecy preserved prestige but prevented transmission. When patrons died or regimes fell, their engineers' secrets vanished with them.

Later guilds continued this pattern. Techniques passed orally to select apprentices. Knowledge was preserved by restriction, not by sharing.

The Shift of Priorities

The disappearance of ancient mechanics was also cultural. As the medieval world shifted toward theology and survival, the spirit of inquiry that had fueled invention dimmed. Devices once built to inspire wonder or express harmony were seen as frivolous or even suspect.

In early Christian Europe, many classical texts were copied for moral instruction rather than practical application. Self moving devices were sometimes viewed with distrust, associated with illusion or pride. The focus shifted from invention to contemplation.

Economic change played a role as well. Times of war and scarcity required bronze and iron for weapons, not for fountains that sang. The engineer, once celebrated in the courts of Alexandria or Baghdad, became a practical craftsman building mills, bridges, and fortifications. Imagination surrendered to necessity.

The Loss of Context

Among the greatest tragedies of this disappearance was not the loss of machines but the loss of understanding. Surviving artifacts puzzled later generations. The Antikythera mechanism, discovered in 1901, revealed astonishing sophistication, yet for centuries similar finds were dismissed as curiosities.

Without the framework of theory, interpretation became impossible. A gear without geometry, a valve without fluid theory, became a relic stripped of meaning. The connection between idea and execution had broken, leaving only mute fragments.

The Interrupted Conversation

For more than a thousand years, humanity lived in a mechanical dark age. The ingenuity of antiquity survived only in rumor. Travelers spoke of "singing birds" in Eastern courts or "self moving statues" in distant temples. These stories sounded like fantasy.

Yet fragments endured. Monasteries guarded partial translations. Arabic libraries preserved diagrams. Legends hinted at lost wonders.

When the Renaissance revived interest in science, scholars found the remnants of ancient mechanics. The rediscovery of Hero's works in the fifteenth century and the translations of Arabic engineering texts restored the long broken conversation. Leonardo da Vinci and his contemporaries joined that dialogue, sketching gears and mechanisms that echoed forgotten masters.

The Cost of Forgetting

History invites a question: What might have been if the thread had never broken? If Su Song's chain drive, Al Jazari's feedback systems, and Hero's steam engines had passed continuously into later centuries, the age of machinery might have dawned centuries earlier. Steam could have powered Roman ships. Mechanical devices might have transformed medieval economies.

Yet the silences shaped history as much as the discoveries. Loss forced rediscovery. Forgetting demanded imagination. The machines vanished, but the impulse to build them endured.

The Echo of Absence

Today, the ruins of ancient engineering still whisper. Archaeologists in Alexandria and Kaifeng uncover fragments of valves and gears. Historians reconstruct manuscripts once thought lost. In their hands, the silence becomes sound again—the breath of air through pipes, the quiet drop of water into a basin.

The disappearance of ancient mechanics reminds us that progress is fragile, dependent on preservation and curiosity. Civilizations fall, but wonder endures.

As the wind sweeps across the buried fountain, grains of sand slip into its empty basin. The air hums faintly, a ghost of music where water once played. Somewhere beneath the stone, bronze pipes still rest, waiting to be found. The machines may be silent, but their story continues—echoing through every modern invention that seeks to give motion back to the still, and meaning back to time.

Chapter 20 Epilogue: The Eternal Mechanism – Humanity's Quest to Breathe Life into Metal

In a glass case at the center of a modern museum, a bronze bird begins to move. Air pushed through hidden pipes whistles through its beak, and a clear note rises, a melody first heard two thousand years ago. Visitors stop where they stand. The sound is soft, almost fragile, yet it awakens something ancient. The bronze glints under the lights, its reflection suspended between the present and the past. For a moment, time folds in on itself, and the bird's song bridges the silence of centuries.

That single sound, mechanical yet alive, captures the essence of humanity's oldest dream, to make the inanimate move and breathe life into matter. Across civilizations and millennia, from the temples of

Alexandria to the courts of Baghdad, from the workshops of Su Song to the studios of Leonardo, people have shared the same impulse, to imitate the living through invention.

The Circle of Creation

Every era rediscovers this desire in its own language. To the Egyptians, motion was sacred, the pulse of the gods within stone and water. To the Greeks, it was reason made visible, the mathematics of the cosmos expressed through mechanism. To the Islamic engineers of Baghdad, it was divine order made known through geometry and balance. To the artisans of Renaissance Italy, it was art itself, a reflection of creation's beauty shaped by human hands.

Today, when a robotic hand flexes or an artificial intelligence composes a melody, we stand within that same continuum. The materials have changed, gears to circuits and levers to code, but the longing is constant. It is the same flame that once powered Hero's temple doors, Al Jazari's fountains, and Su Song's celestial spheres. It is the same spark that transformed curiosity into mechanism and mechanism into meaning.

The Memory of Motion

The rediscovery of ancient machines in the modern era is not just a technical triumph, it is an act of remembrance. Each reconstructed device gives voice to engineers long gone, restoring a conversation interrupted by time. When we rebuild Hero's automata or Su Song's clock tower, we are not simply copying the past, we are completing it.

The Antikythera mechanism, once a corroded mystery, now turns again in digital reconstructions, its gears tracing the paths of planets as they did in the first century BCE. Al Jazari's elephant clock, remade in museums and exhibitions, measures time in honor of a world that once saw no divide between science and art. Each restored device carries the voices of its makers, reminding us that ingenuity is never completely lost, it simply waits to be heard again.

The Divine Reflection

In every age, humanity's machines have acted as mirrors. When priests filled temple pipes to make statues breathe, they mirrored divine creation. When philosophers built automata that imitated thought, they mirrored themselves. The more lifelike the invention, the more clearly it revealed its maker.

This is the enduring paradox of invention: the closer we come to imitating life, the more we confront the nature of life itself. The ancients framed this as theology; we frame it as technology. Yet the questions remain the same. Where does motion end and consciousness begin? Can imitation ever become creation? Is the spark of life mechanical or something beyond measure?

The automaton, ancient or modern, does not answer these questions. It asks them again and again each time it moves.

The Fragile Thread of Genius

For all its brilliance, ancient engineering was fragile. War, conquest, and forgetting buried its marvels beneath centuries of dust. Yet the rediscovery of those marvels affirms another truth, that curiosity cannot be extinguished. Humanity always rebuilds what it admires. When knowledge vanishes, it is recreated from memory, intuition, and the quiet refusal to let wonder die.

The thread that connects Ctesibius to Hero, Al Jazari to Leonardo, and Leonardo to the roboticists of today is not one of unbroken progress but of rediscovery. Each generation finds the same realization, that motion, shaped and purposeful, is a kind of poetry. To make matter move is to enter into a conversation with creation.

The Machine as Memory

In the hum of a modern robot, we can still hear the whisper of ancient fountains. The smooth turn of an industrial gear echoes the rhythm of water dripping in a Greek clepsydra. Even our satellites, circling Earth in silent precision, continue the old dream of aligning human design with celestial motion.

Technology is not a straight march forward but a turning wheel, a cycle of remembering, forgetting, and reimagining. The machines of the past were never truly lost. Their ideas were absorbed into the bloodstream of civilization, waiting for a new pulse to set them in motion.

The Immortal Impulse

The desire to animate the inanimate is more than technical ambition; it is a statement of faith. Whether that faith rests in gods, reason, or creativity itself, it reflects humanity's conviction that the universe is orderly, expressive, and open to understanding. The ancient engineers did not only build mechanisms; they built metaphors for existence.

To make a statue breathe or a clock sing was to express belief in harmony, and to show that metal and water, when arranged with wisdom, could echo the divine.

The Song Continues

In the museum, the bronze bird's song fades, and the visitors applaud softly. A child steps closer to the glass, eyes wide, not in disbelief but in recognition. That wonder is the same wonder that once filled the temples of Alexandria, the gardens of Baghdad, and the workshops of Florence.

The machines have changed, but the impulse behind them has not. It is the same rhythm that drives us to design, to imagine, and to understand, the desire to see life where there is none, and in doing so, to glimpse something eternal.

As the last note of the bronze bird echoes into silence, its song becomes more than a sound. It becomes a memory of everything that has ever moved by human hands and human imagination. The eternal mechanism turns once more, not only in metal and water, but in the heart of every person who believes that wonder is worth creating.

Made in the USA
Monee, IL
04 November 2026

67381303R00089